Topological Objects in Correlated Electronic Systems

Topological Objects in Correlated Electronic Systems

Guest Editors

Serguei Brazovskii
Natasha Kirova

MDPI

Basel • Beijing • Wuhan • Barcelona • Belgrade • Novi Sad • Cluj • Manchester

Guest Editors

Serguei Brazovskii
Laboratory of Theoretical Physics and Statistical Models (LPTMS)
University Paris-Saclay
Orsay
France

Natasha Kirova
Solid State Physics Laboratory (LPS)
University Paris-Saclay
Orsay
France

Editorial Office

MDPI AG
Grosspeteranlage 5
4052 Basel, Switzerland

This is a reprint of the Special Issue, published open access by the journal *Symmetry* (ISSN 2073-8994), freely accessible at: https://www.mdpi.com/journal/symmetry/special_issues/Topological_Objects.

For citation purposes, cite each article independently as indicated on the article page online and as indicated below:

Lastname, A.A.; Lastname, B.B. Article Title. *Journal Name* **Year**, *Volume Number*, Page Range.

ISBN 978-3-7258-5521-6 (Hbk)
ISBN 978-3-7258-5522-3 (PDF)
https://doi.org/10.3390/books978-3-7258-5522-3

Contents

About the Editors

Serguei Brazovskii

Serguei Brazovskii received his PhD and habilitation from the Landau Institute of Theoretical Physics in Russia, where he worked from 1972 to 1998. Since then, he has been employed in France by the CNRS, working at the University Paris-Saclay. S. Brazovskii has also held numerous visiting professorships in Brazil, Croatia, Denmark, Israel, Italy, Japan, Slovenia, and South Korea; in the United States, he served as a consultant at the Los Alamos and Brookhaven National laboratories. Meanwhile, he has served as an editor in several journals and as an organizer of conferences and research schools. S. Brazovskii is the founder and a co-organizer of the series of International School-Colloquiums on Electronic Crystals (ECRYS) and of the series IMPACT (Electronic States and Phases Induced by Electrical or Optical Impacts). He has given invited presentations at numerous international meetings and lectured at many schools. S. Brazovskii is the author of 210 articles published in world-renowned journals. He has collaborated with experimental groups in France, Germany, Japan, Korea, Russia, Slovenia, and the USA. His studies cover a diverse spectrum of condensed matter theory: phase transitions, liquid crystals, strong magnetic and electric fields, collective transformations induced by optical pumping, solitons at microscopic scales and in ensembles, and low-dimensional systems of strongly correlated electrons, such as electronic crystals and charge and spin density waves. His theory of weak crystallization is known as the Landau–Brazovskii transition; the Brazovskii–Kirova model is used to model optics of conducting polymers. He participated in the discovery of electronic ferroelectricity in organic conductors and in establishing the physics of solitons, polarons, and excitons in quasi-one-dimensional systems. His current interests lie mostly in the theory of topological defects, their dynamics, and their collective properties in correlated electronic systems.

Natasha Kirova

Natasha Kirova received her PhD at the Moscow State University and habilitation at the Institute for Chemical Physics in Russia. From 1970 to 2004, she worked in the Moscow center for micro-electronics, while lecturing at the graduate school of microelectronics and in MIREA, Russian Technological University. N. Kirova also held visiting professorships in Brazil, Italy, Japan, South Korea, France, and in the United States, where she served as a consultant at the Los Alamos National laboratory. In 1998, she settled in France, where she works steadily at the Solid State Physics Laboratory (Laboratoire de Physique de Solides) at the University Paris-Saclay. She has served as an organizer of many conferences and research schools, particularly the series of International School-Colloquiums on Electronic Crystals (ECRYS) and of the series IMPACT (Electronic States and Phases Induced by Electrical or Optical Impacts). She gave invited presentations at many international meetings and lectures at several schools. N. Kirova is the author of 147 articles published in world-renowned journals. She collaborates with experimental groups in France, Japan, Russia, South Korea, and the USA. N. Kirova has worked on several subjects of condensed matter theory: domain structure in magnetic materials, strong electric fields in junctions, collective transformations induced by optical pumping, the electronic ferroelectricity, charge and spin density waves, solitons and their superstructures via exact solutions, and many more. In the literature, the Brazovskii–Kirova model of optics in conducting polymers is referred to as a basis for the physics of solitons, polarons, and excitons. Her current interests lie mostly in theory of topological defects, such as various vortices, in coherent electronic states, and in theory of the optically induced dynamical excitonic state.

symmetry

MDPI

Editorial

Topological Objects in Ordered Electronic Systems

Serguei Brazovskii [1,*] and Natasha Kirova [2,3,*]

[1] CNRS, Laboratoire de Physique Théorique et Modèles Statistiques, University Paris-Saclay, 91405 Orsay, France
[2] CNRS, Laboratoire de Physique des Solides, University Paris-Saclay, 91405 Orsay, France
[3] Russian Quantum Center, 143025 Skolkovo, Russia
* Correspondence: serguei.brazovski@universite-paris-saclay.fr (S.B.); natacha.kirova@universite-paris-saclay.fr (N.K.)

Most of correlated electronic systems possess ground states with broken crystal symmetries. These include superconductivity, spin orderings, a vast family of electronic crystals (including charge-/spin-density waves, Wigner crystals, arrays of stripes, charge ordering, and electronic ferroelectrics), and other translationally periodic states like super-structures in spin systems, spin-polarized density waves, and superconductors. Ground-state degeneracy allows for topologically nontrivial configurations connecting equivalent, whilst different, states. These "topological defects" include extended objects such as plane domain walls, lines of dislocations or phase vortices, various solitons or skyrmions as local microscopic objects, and transient processes—referred to as instantons—such as phase slips as space–time vortices. Embedded or transient topologically nontrivial configurations are readily induced by doping or optical pumping, by electric or magnetic fields, and under stresses or sliding. The Special Issue of the MDPI journal *Symmetry* and this corresponding e-book address these phenomena, as well as other topology-related electronic properties. This editorial article offers a schematic introduction to the field, placing emphasis on microscopic solitons and density waves, and provides a short summary of the content of this Special Issue.

1. Topologically Nontrivial Configurations: From Condensed Matter via the Solid State to Electronic Realizations

Embedded or transient topologically nontrivial configurations are common among symmetry-broken ground states [1]. State degeneracy tolerates the formation of various configurations, commonly called topological defects, connecting equivalent but different ground states

The importance of topology-related phenomena in condensed matter physics has been recognized in the last decade. For example, in 2014, the Lars Onsager Prize was awarded to V.P. Mineev and G.E. Volovik "for or their contribution to a comprehensive classification of topological defects in condensed matter phases with broken symmetry". Then, in 2016, the Nobel Prize in Physics was awarded to D.J. Thouless, F.D. Haldane, and J.M Kosterlitz, more generally "for theoretical discoveries of topological phase transitions and topological phases of matter".

The classification of topological defects and questions on their allowance, stability, and protection primarily depend on the dimensions of the degeneracy manifold of their order parameters with respect to the dimensions of the space in which they are embedded [2,3]. Topological defects include macroscopic extended objects (domain walls, dislocation lines, and vorticity lines of phases or directors), and microscopic ones, including various solitons [4] and instantons for related transient processes.

The numerous existing or discussed realizations of topological defects involve particle physics [5], cosmology [6,7], quantum liquids [8,9] and gases, optical condensates [10], cold atoms [11], liquid crystals [12], conjugated polymers [13], and other quasi-one-dimensional conductors and biological macro-molecules, etc.

Solid states reveal topological defects, such as as domain walls or discommensurations in superstructures ([14] in this Special Issue), current vortices and phase slips in superconductors [15], displacement vortices as dislocations [16], vortices [17,18], and phase slips [19] in sliding superstructures, walls, and skyrmions in magnetic media [20–22].

Topological defects appear in the frame of strongly correlated electronic systems which typically show various types of symmetry breaking, giving rise to degenerate ground states. Among them, the vast family of electronic crystals includes charge- and spin-density waves; Wigner crystals; arrays of stripes [23–25], which will be discussed in this Special Issue; charge ordering; electronic ferroelectrics; and spin superstructures under magnetic fields. The latter include the spin-Peierls state [26], spin-polarized charge-density waves, and the FFLO state [27–29] in superconductors. Some of these superlattices, particularly incommensurate density waves [23,30–32] and Wigner crystals [23], are able to maintain the collective electric current by means of sliding (the so called Frölich conductivity [23,30,33]). For these incommensurate electronic crystals, their number of unit cells is not fixed which allows to absorb excess electrons into the extended ground state. This exchange among normal and condensed charge carriers requires for steps involving topologically nontrivial objects, like amplitude and phase solitons; phase vortices – dislocations, proceeding via transient processes of phase slips); and space–time vortices [15,17,19]. All of these result in a rich complex of nonlinear and nonstationary behavior with vast experimental observations. Here, topological defects are crucially necessary for the conversion between normal and collective currents, and they also appear in depinning processes to initiate sliding in the presence of host defects and constraints.

Among electronic crystals, the most frequently studied ordered states are the charge- and spin-density waves, which are ubiquitous in quasi-one-dimensional systems [30–32]. They demonstrate spectacular nonlinear conduction through collective sliding and lability to the electric field and the current injection. Static and transient topological defects emerge necessarily to maintain these phenomena; dislocations as space vortices and space–time vortices are known as instantons or phase-slip centers. Dislocations [18] are built-in statically under a transverse electric field [34]; their sweeping provides conversion among the normal carriers and the condensate [35,36], which ensures the onset of collective sliding. A special realization in a high magnetic field [37], when the density wave is driven by the Hall voltage originated by the current of quantized normal carriers, reveals the dynamical vorticity that serves to annihilate compensating normal and collective currents.

In charge-density waves, topological defects appear under stresses coming from various sources: surface mismatches of periodicity [38] or injection ([14,39] in this Special Issue), proximity to commensurability ([40] in this Special Issue), a constraint geometry [41]; and an imbalance of normal and collective currents near junctions in the sliding regime [35,36]. The stress can easily exceed a plastic threshold, leading to the appearance of topological defects. For commensurate or close-to-commensurability density waves, the related strains can leave particular fingerprints, like a solitonic lattice or a system of random solitons, which allows for the identification of underlying structures via the STM [42,43] or by space-resolved X-ray scattering ([35,36,40] in this Special Issue).

Finally, spin-density waves, with their rich multiplicative order parameter, reveal complex objects with half-integer topologically bound vorticities in phases and directional degrees of freedom ([44] in this Special Issue).

2. Microscopic Solitons in (Quasi)-One-Dimensional Electronic Systems

The popular notion of "topological solitons" is a shortening for "topologically nontrivial solitons" which still need to be more precise, such as "topologically stable ..." or "topologically protected ...". The "topological stability" is based on conservation of "topological charges" and the related "irreducibility of trajectories" connecting different states of a systems, see reviews e.g., [2,3].

Here, in applications to electronic systems, following only the minimal feature of a broadly spanning trajectory with no respect to its reducibility , we imply a broader, while less precise, definition of a "topologically nontrivial object", as a local configuration exploring manifold degenerate ground states, connecting different equivalent ones. Their configurations may not be topologically stable, thus allowing for their transmutations with trivial electronic excitations. Such solitons appear or are preserved because they are energetically preferable, and/or because the total electric charge or spin, which these solitons carry, is preserved or monitored externally.

In one-dimensional systems, topological solitons become truly microscopic objects, carrying energy, charge, and other quantum numbers at a single-electron scale. As microscopic quasi-particles, they may become the lowest energy excitations and take over the role of conventional electrons in transport or optical properties (see short reviews [45,46]). Since the solitons possess similar quantum numbers such as the charge or the spin, their stable ensembles can be controllably created, maintained, and observed. Such ensembles can experience a sequence of phase transitions, accompanied by the formation of structures at an increasing scale: from individual solitons, via their microscopic complexes and growing aggregates, to macroscopic domain walls and stripes.

The role of solitons in electronic properties was anticipated in theories since the mid 1970s (see short reviews [45,46]). The common double degeneracy of systems with dimerization, like (sin-)Peierls states or the Mott state with a charge ordering), gives rise to solitons as kinks of the scalar order parameter A. The continuous degeneracy of the complex order parameter $A\exp(i\theta)$ (superconductors or charge-density waves) gives rise to phase vortices, amplitudes solitons, and their combinations. These degrees of freedom can be controlled or accessed independently via either the spin polarization or the charge doping.

There is a vast amount of experimental evidence on the existence of microscopic solitons and their determining role in electronic processes of quasi-one-dimensional electronic crystals. Different types of solitons appear in experiments. These include the amplitude king which can be charged spinless or neutral spin carrying, and the "holon" and the "spinon", which are general terms used for strongly correlated electronic systems. There are also the polarization kinks carrying a fractional charge ([47] in this Special Issue), and the topologically bound charge–spin soliton. Solitons were firstly accessed in experiments on conducting polymers in the early 1980s [48]. They received renewed attention in the early 2000s from discoveries of ferroelectric charge ordering (see a review [49]) in organic conductors, from access via nano-scale tunneling experiments [41] in materials with charge-density waves, and from the optics of conducting polymers [50]. Today, various solitons appear in conductivity, tunneling spectroscopy, and optical absorption. Instantons, the corresponding dynamical processes, are responsible for subgap transitions leading to a pseudogap formation. The ferroelectric charge ordering in organic conductors provides access to several types of solitons ([47,49,51] in this Special Issue) observed in conductivity (holons), NMR (spinons), permittivity (polar kinks), as bound pairs in optics, and also compound charge–spin solitons in cases of combined symmetry breaking in polymers [50] and in charge-transfer molecular crystals ([47,51] in this Special Issue). In charge-density waves, individual solitons as the amplitude kinks, namely the spinons, have been visually captured in STM experiments [42,43]; notably, they have also been captured in diverse local probes in magnetic systems ([52] in this Special Issue). The subgap spectra of the coherent internal tunneling [41] recover the solitons as instantons.

Beyond the exceptional case of a truly one-dimensional (1D) system [43], experimentally, solitons have been observed or looked for in quasi-one-dimensional systems within low-temperature phases with long-range orders. Here, commuting between degenerate minima at only one chain would lead to a loss of the inter-chain-ordering energy proportional to the distance along the chain until the next soliton or a boundary. This energy dominates at long distances, even if it can be unimportant locally for a weak inter-chain interaction, which gives rise to the confinement of solitons (see [45,53] in this Special Issue). These interactions can appear already in some specific 1D systems where the ground-state degeneracy is not exact, so the soliton connects the true and false vacuums, losing the confinement energy.

The effect of confinement is omnipresent at higher dimensions where the inter-chain interactions, which are responsible for establishing the long-range 2D or 3D ordering, lift the degeneracy locally. In cases of discrete symmetries, the solitons are bound in topologically trivial pairs with an option for a subsequent phase transition to form cross-sample domain walls. In cases of continuous symmetries, the gapless mode can cure solve the interruption from the amplitude kink, which allows for individual solitons to exist in the low-temperature phases with long-range ordered states. The solitons adapt by forming topologically bound combined complexes with half-integer vortices of gapless modes: π-rotons [45,46]. In cases of repulsing and attracting electronic interactions correspondingly, this results in spin- or charge-roton configurations with charge- or spin-amplitude solitons localized in the core.

For the ensemble with a finite density of solitons, the confinement forces lead to a sequence of phase transitions ([53] in this Special Issue). The higher-temperature transition forces the confinement of solitons into topologically bound complexes: pairs of kinks or the amplitude solitons dressed by exotic half-integer vortices. At a second, lower temperature transition, the solitons aggregate into rods of bi-kinks or into walls of amplitude solitons terminated by rings of half-integer vortices [29,45,46,53]. As temperature lowers , the walls multiply, passing sequentially across the sample.

3. Special Issue Summary

This Special Issue features nine articles (including this introduction) which largely, though not exclusively, cover contemporary studies of topological objects in electronic, and related, systems.

The article by J. Tranquada [25] describes the concept of topological doping realized in high-Tc cuprate superconductors. Hole doping into a correlated antiferromagnet leads to charge stripes that separate antiferromagnetic spin stripes from opposite phases. The anti-phase Josephson coupling across the spin stripes can lead to a pair-density-wave order in which the broken translation symmetry of the superconducting wave function is accommodated by pairs with finite momentum.

The article by A. Kranjec et al. [14] describes studies in the dynamics of electronic dislocations and discommensurations in a Wigner crystal state. The studies exploit the scanning probe microscopy of an ensemble of structurally ordered polarons perturbed by optical pulses or a local charge injection. The experiments were performed on a layered compound which exhibits the Wigner crystal, displaying discommensurations and domain patterns when an additional charge is injected, either through contact or by photoexcitation. The domain walls and their crossings display topologically metastable entangled structures. The studies demonstrate the significance of topological protection at the microscopic level: the topologically trivial defects are rapidly annihilated pair-density-wave order with respect to the (meta-)stable non-trivial defects.

The article by J. Seidel [52] presents a scanning probe microscopy investigation of topological defects in magnetically and/or ferroelectrically ordered media. The article offers a comprehensive comparison of various versions of the techniques, from the more common STM, AFM, and SQUID to less known ones. The examples of nano-scale topological defects in the study span from straight domain walls to nontrivial monopoles and skyrmions.

The article by L. Vigliotti et al. [54] investigates bound states in a one-dimensional topological superconductor, addressing the dilemma of competition between the cases of Majorana- and Tamm-type edge states. This question is important in view of envisaged applications in topological quantum computation, which require the engineering of non-Abelian Majorana zero modes, the presence of which can be misleading due to the appearance of Tamm- or Andreev-bound states. The authors designed a model in which both Majorana- and Tamm-bound states can be present and compete. The model is a finite-size, one-dimensional topological superconductor in the presence of a competing normal gapping mechanism arising from a position-dependent potential, akin to a CDW, in which the phase can be monitored.

The article by K. Sunami [47] et al. is devoted to studies of solitons appearing in a quasi-one-dimensional ferroelectric-conducting material, composed of stacks of alternating donor and acceptor molecules. The dimerizational symmetry breaking which accompanies the neutral—ionic phase transition gives rise to solitonic topological excitations as mobile boundaries between alternating ferroelectric domains. These solitons are expected to carry fractional charges, and may also carry the electronic spins that lead to the anomalous charge transport and the spin response. The article reviews related properties, which are studied by using a combination of NMR, NQR, and electrical resistivity measurements, and recalls the underlying theoretical concepts.

The article by D. Le Bolloc'h et al. [40] describes the application of space-resolved coherent X-ray diffraction in studies of solitonic lattices in sliding charge-density waves. The main features emerging from the local probe experiments include the influence of charge-density wave pinning on the sample surfaces and the propagation of periodic phase defects, such as charge solitons, across the entire sample. Corresponding numerical modeling is also presented.

The article by P. Karpov et al. [53] presents numerical simulations of the pattern formation and aggregation across phase transition in ensembles of solitons in a quasi-one-dimensional system. The long-range ordering enforced by the inter-chain coupling imposes super-long-range confinement forces upon the solitons, leading to a sequence of phase transitions in their ensembles. The higher-temperature transition enforces the confinement of solitons into topologically bound complexes, comprising pairs of kinks or the amplitude solitons dressed by half-integer vortices. At a second, lower-temperature transition, the solitons aggregate into rods of bi-kinks or into walls of amplitude solitons terminated by rings of half-integer vortices. The efficient Monte Carlo algorithm was employed, allowing to extend them to the three-dimensional case and to include the long-range Coulomb interactions.

The article by N. Kirova et al. [44] presents the phenomenological theory and numerical simulations of dynamical electronic vortices in charge- and spin-density waves. Their collective sliding requires the emergence of static and transient topological defects: there are dislocations as space vortices and space–time vortices known as phase-slip centers, which are a kind of instantons. The rich order parameter of spin-density waves reveals complex objects with half-integer topologically bound vorticities in charge- and spin degrees of freedom. The presented modeling is based upon numerical solutions to partial differential equations for the dissipative dynamics. It takes into account the complex order parameter, the self-consistent electric field, and the normal carriers. The traditional time-dependent Ginzburg–Landau approach, which is shown to be contradictory with respect to the charge conservation, was generalized, allowing the authors to treat the intrinsic normal carriers consistently.

In conclusion, we believe that this Special Issue and the subsequent e-book will illustrate the vitality of the field and the diversity of its subjects. We appreciate the efforts of the authors and help from referees, all of whose contributions have made this Special Issue a true success. We acknowledge the Editorial Office for supporting this Special Issue and the edited e-book.

Author Contributions: All authors have equally contributed to the article. All authors have read and agreed to the published version of the manuscript.

Funding: This research received no external funding.

Conflicts of Interest: The authors declare no conflicts of interest.

References

1. Bunkov, Y.M.; Godfrin, H. (Eds.) *Topological Defects and the Non-Equilibrium Dynamics of Symmetry-Breaking Phase Transitions;* NATO ASI Series; Kluwer Academic Publishers: Amsterdam, The Netherlands, 2000.
2. Mermin, N.D. The topological theory of defects in ordered media. *Rev. Mod. Phys.* **1979**, *51*, 591–648. [CrossRef]
3. Mineev V.P. *Topologically Stable Defects and Solitons in Ordered Media;* Harwood Academic Publishers: New York, NY, USA, 1998.
4. Manton, N.; Sutcliffe, P. *Topological Solitons;* Cambridge University Press: Cambridge, UK, 2004.
5. Skyrme, T.H.R. A non-linear field theory. *Proc. R. Soc. Lond. Ser. A* **1961**, *260*, 127–138.
6. Hindmarsh, M.W.; Kibble, T.B.W. Cosmic strings. *Rep. Prog. Phys.* **1995**, *58*, 477–561. [CrossRef]
7. Vilenkin, A.; Shellard, E.P.S. *Cosmic Strings and Other Topological Defects;* Cambridge University Press: Cambridge, UK, 2000.
8. Eltsov, V.B.; Krusius, M.; Volovik, G.E. Vortex formation and dynamics in superfluid ^{3}He and analogies in quantum field theory. *Prog. Low Temp. Phys.* **2005**, *15*, 1–137.
9. Volovik, G. Monopoles and fractional vortices in chiral superconductors. *Proc. Natl. Acad. Sci. USA* **2000**, *97*, 2431–2436. [CrossRef] [PubMed]
10. Berloff, N.G.; Keeling, J. Universality in Modelling Non-equilibrium Pattern Formation in Polariton Condensates. *Physics of Quantum Fluids: New Trends and Hot Topics in Atomic and Polariton Condensates;* Springer Series 177; Bramati, A., Modugno, M., Eds.; Springer: Berlin/Heidelberg, Germany, 2013; pp. 19–38.
11. Neely, T.W.; Scherer, D.R.; Bradley, A.S.; Davis, M.J.; Anderson, B.P. Spontaneous vortices in the formation of Bose-Einstein condensates. *Nature* **2008**, *455*, 948–951.
12. Chuang, I.; Durrer, R.; Turok, N.; Yurke, B. Cosmology in the laboratory: Defect dynamics in liquid crystals. *Science* **1991**, *251*, 1336–1342. [CrossRef] [PubMed]
13. Yu, L. *Solitons and Polarons in Conducting Polymers;* World Scientific Publishing Co.: New York, NY, USA, 1988.
14. Kranjec, A.; Karpov, P.; Vaskivskyi, Y.; Vodeb, J.; Gerasimenko, Y.; Mihailovic, D. Electronic Dislocation Dynamics in Metastable Wigner Crystal States. *Symmetry* **2022**, *14*, 926. [CrossRef]
15. Ivlev, B.I.; Kopnin, N.B. Electric currents and resistive states in thin superconductors. *Adv. Phys.* **1984**, *33*, 47–114. [CrossRef]
16. Landau, L.D.; Pitaevskii, L.P.; Kosevich, A.M.; Lifshitz, E.M. *Theory of Elasticity;* Pergamon: Oxford, UK, 1984.
17. Ong, N.P.; Maki, K. Generation of charge-density-wave conduction noise by moving phase vortices. *Phys. Rev. B* **1985**, *32*, 6582–6590. [CrossRef] [PubMed]
18. Feinberg, D.; Friedel, J. Elastic and plastic deformations of charge density waves. *J. Phys.* **1988**, *49*, 485–496. [CrossRef]
19. Gor'kov, L.P. Phase slipping processes and generation of narrow-band oscillations by charge density waves. In *Charge Density Waves in Solids;* Gor'kov, L., Grüner, G., Eds.; Elsevier Science Publisher: Amsterdam, The Netherlands, 1990; pp. 403–424.
20. Kosevich, A.M.; Ivanov, B.A.; Kovalev, A.S. Magnetic Solitons. *Phys. Rep.* **1990**, *194*, 117–238. [CrossRef]
21. Nagaosa, N.; Tokura, Y. Topological properties and dynamics of magnetic skyrmions. *Nat. Nanotechnol.* **2013**, *8*, 899–911. [CrossRef] [PubMed]
22. Seidel, J. (Ed.) *Topological Structures in Ferroic Materials: Domain Walls, Vortices and Skyrmions;* Springer Series in Materials Science; Springer: Berlin/Heidelberg, Germany, 2016.
23. Monceau, P. Electronic crystals: An experimental overview. *Adv. Phys.* **2012**, *61*, 325–581. [CrossRef]
24. Brazovskii, S.; Kirova, N. Proceedings of Iternational Conference ECRYS 2014. *Phys. Condens. Matter.* **2015**, *460*, 1–268. [CrossRef]
25. Tranquada, J.M. Topological Doping and Superconductivity in Cuprates: An Experimental Perspective. *Symmetry* **2021**, *13*, 236. [CrossRef]

26. Horvatic, M.; Fagot-Revurat, Y.; Berthier, C.; Dhalenne, G.; Revcolevschi, A. NMR Imaging of the Soliton Lattice Profile in the Spin-Peierls Compound $CuGeO_3$. *Phys. Rev. Lett.* **1999**, *83*, 420–423. [CrossRef]
27. Machida K.; Nakanishi, H. Superconductivity under a ferromagnetic molecular field. *Phys. Rev. B* **1984**, *30*, 122–133. [CrossRef]
28. Buzdin A.I.; Tugushev, V.V. Phase diagrams of electronic and superconductlng transitions to soliton lattice states. *Sov. Phys. JETP* **1983**, *58*, 428–433.
29. Radzihovsky, L.L.; Vishwanath, A. Quantum Liquid Crystals in an Imbalanced Fermi Gas: Fluctuations and Fractional Vortices in Larkin-Ovchinnikov States. *Phys. Rev. Lett.* **2009**, *103*, 010404. [CrossRef] [PubMed]
30. Grüner, G. *Density Waves in Solids*; CRC Press: Boca Raton, FL, USA, 2018.
31. Gor'kov, L.; Grüner, G. (Eds.) *Charge Density Waves in Solids*; Elsevier Science Publisher: Amsterdam, The Netherlands, 1990.
32. Grüner, G. The Dynamics of Spin Density Waves. *Rev. Mod. Phys.* **1994**, *66*, 1–24. [CrossRef]
33. Bardeen, J. Superconductivity and Other Macroscopic Quantum Phenomena. *Phys. Today* **1990**, *43*, 25–33. [CrossRef]
34. Brazovskii, S.; Matveenko, S. The charge density wave structure near a side metal contact. *Sov. Phys. JETP* **1992**, *74*, 864–869. [CrossRef]
35. Requardt, H.; Nad, F.; Monceau, P.; Currat, R.; Lorenzo, J.E.; Brazovski, S.; Kirova, N.; Grubel, G.; Vettier, C. Direct observation of CDW current conversion by spatially-resolved synchrotron X-ray studies in $NbSe_3$. *Phys. Rev. Lett.* **1998**, *80*, 5631–5634. [CrossRef]
36. Lemay, S.G.; de Lind van Wijngaarden, M.C.; Adelman, T.L.; Thorne, R.E. Spatial distribution of charge-density-wave phase slip in NbSe3. *Phys. Rev. B* **1998**, *57*, 12781–12791. [CrossRef]
37. Orlov, A.P.; Sinchenko, A.A.; Monceau, P.; Brazovskii, S.; Latyshev, Y.I. Hall voltage drives pulsing counter-currents of the sliding charge density wave and of quantized normal carriers at self-filled Landau levels. *NPJ Quantum Mater.* **2017**, *2*, 61. [CrossRef]
38. Kirova, N.; Brazovskii, S. Scattering Profile from a Random Distribution of Dislocations in a Charge Density Wave. *Ann. Phys.* **2022**, *447*, 169130. [CrossRef]
39. Cho, D.; Cheon, S.; Kim, K.-S.; Lee, S.-H.; Cho, Y.-H.; Cheong, S.-W.; Yeom, H.W. Nanoscale Manipulation of the Mott Insulating State Coupled to Charge Order in 1*T*-TaS_2. *Nat. Commun.* **2016**, *7*, 10453. [CrossRef] [PubMed]
40. Le Bolloc'h, D.; Bellec, E.; Kirova, N.; Jacques, V.L.R. Tracking Defects of Electronic Crystals by Coherent X-ray Diffraction. *Symmetry* **2023**, *15*, 1449. [CrossRef]
41. Latyshev, Y.I.; Monceau, P.; Brazovskii, S.; Orlov, A.P.; Fournier, T. Subgap Collective Tunneling and its Staircase Structure in Charge Density Waves. *Phys. Rev. Lett.* **2006**, *96*, 116402. [CrossRef] [PubMed]
42. Brazovskii, S.; Brun, C.; Wang, Z.-Z.; Monceau, P. Scanning Tunneling Microscope Imaging of Single-Electron Solitons in a Material with Incommensurate Charge Density Waves. *Phys. Rev. Lett.* **2012**, *108*, 096801. [CrossRef] [PubMed]
43. Kim, T.H.; Yeom, H.W. Topological Solitons versus Nonsolitonic Phase Defects in a Quasi-One-Dimensional Charge-Density Wave. *Phys. Rev. Lett.* **2012**, *109*, 246802. [CrossRef] [PubMed]
44. Kirova, N.; Brazovskii, S. Simulations of Dynamical Electronic Vortices in Charge and Spin Density Waves. *Symmetry* **2023**, *15*, 915. [CrossRef]
45. Brazovskii, S. New Routes to Solitons in Quasi One-Dimensional Conductors. *Solid State Sci.* **2008**, *10*, 1786–1789. [CrossRef]
46. Brazovskii, S. Microscopic Solitons in Correlated Electronic Systems: Theory versus Experiment. In *Advances in Theoretical Physics: Landau Memorial Conference*; AIP Conference Proceedings; American Institute of Physics: College Park, MD, USA, 2009; Volume 1134, pp. 74–82.
47. Sunami, K.; Takehara, R.; Miyagawa, K.; Okamoto, H.; Kanoda, K. Topological Excitations in Neutral–Ionic Transition Systems. *Symmetry* **2022**, *14*, 925. [CrossRef]
48. Heeger, A.J.; Kivelson, S.; Schrieffer, R.; Su, W.-P. Solitons in Conducting Polymers. *Rev. Mod. Phys.* **1998**, *60*, 781–858. [CrossRef]
49. Brazovskii, S. Ferroelectricity and Charge Ordering in Quasi One-Dimensional Organic Conductors. In *Physics of Organic Superconductors and Conductors*; Lebed, A.G., Ed.; Springer Series in Materials Sciences, 110; Springer: New York, NY, USA, 2008; pp. 313–356.
50. Korovyanko, O.J.; Gontia, I.I.; Vardeny, Z.V.; Masuda, T.; Yoshino, K. Ultrafast Dynamics of Excitons and Solitons in Disubstituted Polyacetylene. *Phys. Rev. B* **2003**, *67*, 035114. [CrossRef]
51. Kagawa, F.; Horiuchi, S.; Matsui, H.; Kumai, R.; Onose, Y.; Hasegawa, T. Tokura, Y. Electric-Field Control of Solitons in a Ferroelectric Organic Charge-Transfer Salt. *Phys. Rev. Lett.* **2010**, *104*, 227602. [CrossRef] [PubMed]
52. Seidel, J. Scanning Probe Microscopy Investigation of Topological Defects. *Symmetry* **2022**, *14*, 1098. [CrossRef]
53. Karpov, P.; Brazovskii, S. Pattern Formation and Aggregation in Ensembles of Solitons in Quasi One-Dimensional Electronic Systems. *Symmetry* **2022**, *14*, 972. [CrossRef]
54. Vigliotti, L.; Cavaliere, F.; Carrega, M.; Ziani, N.T. Assessing Bound States in a One-Dimensional Topological Superconductor: Majorana versus Tamm. *Symmetry* **2021**, *13*, 1100. [CrossRef]

symmetry

MDPI

Review

Topological Doping and Superconductivity in Cuprates: An Experimental Perspective

John M. Tranquada

Brookhaven National Laboratory, Condensed Matter Physics & Materials Science Division, Upton, NY 11973, USA; jtran@bnl.gov

Abstract: Hole doping into a correlated antiferromagnet leads to topological stripe correlations, involving charge stripes that separate antiferromagnetic spin stripes of opposite phases. The topological spin stripe order causes the spin degrees of freedom within the charge stripes to feel a geometric frustration with their environment. In the case of cuprates, where the charge stripes have the character of a hole-doped two-leg spin ladder, with corresponding pairing correlations, anti-phase Josephson coupling across the spin stripes can lead to a pair-density-wave order in which the broken translation symmetry of the superconducting wave function is accommodated by pairs with finite momentum. This scenario is now experimentally verified by recently reported measurements on $La_{2-x}Ba_xCuO_4$ with $x = 1/8$. While pair-density-wave order is not common as a cuprate ground state, it provides a basis for understanding the uniform d-wave order that is more typical in superconducting cuprates.

Keywords: superconductivity; cuprates; stripe order; pair-density-wave order

1. Introduction

Charge order has now been observed in virtually all hole-doped cuprate superconductor families [1–3]. In 214 cuprates, such as $La_{2-x}Sr_xCuO_4$ (LSCO) and $La_{2-x}Ba_xCuO_4$ (LBCO), the charge-stripe order is generally accompanied by spin-stripe order [4–8], as originally observed in Nd-doped $La_{2-x}Sr_xCuO_4$ [9,10]; each of these orders breaks the translation symmetry of the square-lattice CuO_2 planes. In a 1996 paper, Kivelson and Emery [11] pointed out the topological character of the combined spin and charge stripe orders. This corresponds to the fact that the period of the spin-stripe order is twice that of the charge-stripe order, as the antiferromagnetic phase flips by π across each charge stripe, as illustrated in Figure 1.

The topological character of stripes in cuprates is distinct from that of the topological insulators that have dominated attention more recently [12,13]. In the latter case, the focus is on Bloch states in which spin-orbit effects play a special role. In cuprates, in contrast, the effects of strong onsite Coulomb repulsion among Cu $3d$ electrons tend to make Bloch states of questionable relevance. In a parent compound, such as La_2CuO_4, one has a single unpaired Cu $3d_{x^2-y^2}$ electron on each Cu atom that acts as a local moment, with neighboring moments coupled antiferromagnetically by superexchange J, a local interaction. While the electronic band gap has a charge-transfer character due to O $2p$ states that lie between the lower and upper Hubbard bands associated with the Cu $3d_{x^2-y^2}$ orbital, it is the locally antiferromagnetic (AF) environment that limits the motion of doped holes.

It has taken quite some time to appreciate the significance of the topological order associated with spin stripes. Experimentally, the same antiphase relationship of spin stripes seen in superconducting cuprates also occurs in the case of insulating behavior in $La_{2-x}Sr_xNiO_4$ [14] and in $La_{2-x}Sr_xCuO_4$ with $0.02 \lesssim x \lesssim 0.05$ [15] (where the stripes run diagonally with respect to the Ni–O or Cu–O bonds). A theoretical analysis of interaction requirements for topological doping came to no firm conclusions [16]. Antiphase spin stripes have been obtained from many different approaches: from Hartree–Fock calculations on the Hubbard model [17], from effective models that include long-range Coulomb interactions [18], and from advanced variational and quantum Monte Carlo evaluations of the t-J [19] or Hubbard model [20,21].

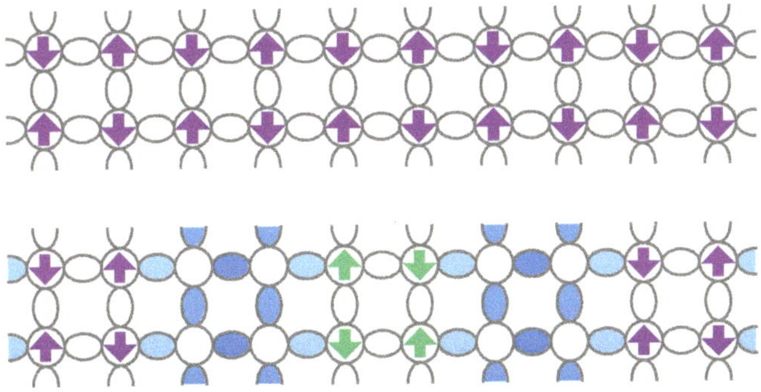

Figure 1. Upper panel indicates the antiferromagnetic order of the undoped CuO_2 planes, with spin direction (arrows) indicated on Cu atoms (circles), separated by O atoms (ellipses). Lower panel shows the spin configuration in the stripe-ordered phase at a doped-hole concentration of $p = 1/8$, with doped hole density indicated by blue shading; antiphase spin stripe indicated in green.

I argued recently [22] that the key feature of topological doping is that the spin degrees of freedom within the charge stripes feel a geometric frustration of their interactions with the neighboring spin stripes. This allows the charge stripes to develop quasi-one-dimensional spin correlations. In the case of cuprates with bond-parallel stripes, the charge stripes may be viewed as hole-doped, two-leg, spin $S = 1/2$ ladders, which are established to have strong superconducting correlations [23,24]. This is a variation on the original proposal of superconducting charge stripes by Emery, Kivelson, and Zachar [25], who pointed out that a spin gap in a one-dimensional (1D) electron gas acts as a pairing amplitude; the difference is that they assumed that the spin gap would be transferred from the neighboring spin stripes, in which case one would never achieve superconductivity when the spin-stripe order is present. The advantage of the doped two-leg spin ladder is that it comes with its own spin gap.

To obtain superconducting order in the CuO_2 planes, it is necessary to establish phase coherence, via Josephson coupling, between neighboring charge stripes [25]. Because of the conflict between local AF order and hole motion, this needs to be antiphase superconducting order, resulting in a pair-density-wave (PDW) state [26,27]. PDW order was initially proposed [28,29] to explain the experimental observation of two-dimensional superconductivity in CuO_2 layers [30], with frustration of the usual Josephson coupling between planes [31].

While the initial case for PDW order was circumstantial, direct phase-sensitive evidence of PDW order in LBCO $x = 1/8$ has now been reported [32]. This result is consistent with measurements of the Hall effect in high magnetic fields along the c-axis that suggest that the holes in the charge stripes remain paired, even in the absence of superconducting order [33]. Hence, there is now a solid case that charge stripes in cuprates are essential to pairing.

Of course, the superconducting ground state of most cuprates is the spatially uniform d wave, not PDW. This is still compatible with pairing correlations developing within charge stripes, but it requires disordered spin stripes with an energy gap [22,34]; uniform phase coherence can only be achieved at energies below the spin gap. The antiphase spin stripes play a critical role for the superconducting order: they either need to be ordered to allow the PDW phase order to be established, or gapped to enable a spatially uniform superconducting order. As a consequence, uniform superconductivity will not coexist with a PDW ground state. On the other hand, defects that require the superconducting order parameter to locally go to zero can favor the local PDW order without a spin order, as seen in studies by scanning tunneling spectroscopy [35,36].

In the following, I fill in details that provide support for the story laid out above.

2. Stripe Order and Decoupling of Spin Excitations

The holes doped into the CuO_2 planes tend to go into O $2p$ states [37]. As pointed out by Emery and Reiter [38], if one could localize a single hole, it would cause the neighboring Cu moments to be parallel, which frustrates the AF order of the undoped system. In fact, it takes very few holes to kill the AF order. In LSCO, the commensurate AF order is gone by $p = 0.02$, and even before that, one has phase separation at low temperature [39]. This transition occurs at a hole density that is 20 times smaller than the limit for percolation due to substitution of nonmagnetic ions, as verified in LSCO with nonmagnetic Zn and Mg substitution for Cu [40].

Initially, the holes form diagonal stripes [15] and the system is insulating. This is similar to $La_{2-x}Sr_xNiO_4$ (LSNO) and $La_2NiO_{4+\delta}$ [41–43]. The case of LSNO with $x = 1/3$ is of particular interest. Neutron scattering measurements of magnetic scattering are presented in Figure 2a; these can be understood in terms of the stripe order illustrated in Figure 3. Note that in contrast to the spin $S = 1/2$ of Cu^{2+}, the Ni^{2+} sites have $S = 1$. The Ni moments on the spin stripes order [44] and exhibit well-defined spin waves [45,46]. Within the charge stripes, there is one hole per Ni site; a low-spin hybridization is expected to leave a net $S = 1/2$ per Ni site along a charge stripe. The interaction of each such moment with the neighboring spin stripes is geometrically frustrated. It is still possible for the reduced Ni moments to couple antiferromagnetically along a charge stripe. For such a decoupled 1D spin chain, one would expect to see no order but spin excitations that disperse only along the stripe direction. Just such 1D spin excitations were first identified by Boothroyd et al. [47]; the role of the decoupling of interactions due to such site-centered charge stripes between antiphase spin stripes was recognized and confirmed in [48].

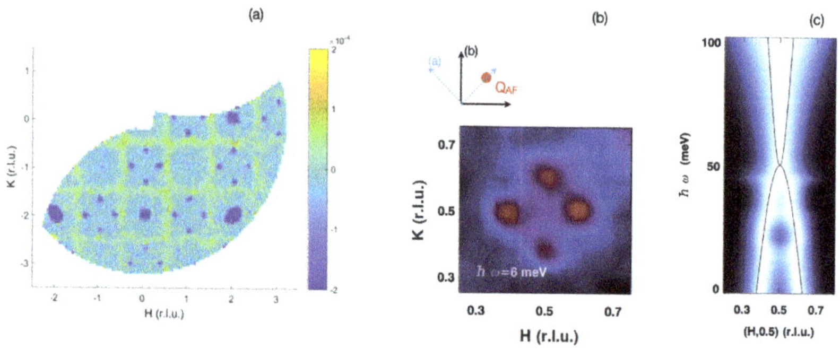

Figure 2. (**a**) Difference in neutron scattering intensity measured at 5 K and 70 K for $\hbar\omega = 3 \pm 1$ meV in $La_{2-x}Sr_xNiO_4$ with $x = 1/3$. Dark blue points at positions of the type $(1 \pm \frac{1}{3}, 0, 0)$ and $(1, \pm\frac{1}{3}, 0)$ correspond to spin waves associated with the spin stripe order, where the AF wave vector, $\mathbf{Q}_{AF}$, is $(1, 0, 0)$. Yellow lines correspond to cuts through 2D planes of scattering from 1D spin correlations in charge stripes. Note that twinning causes the measurement to include scattering from stripe domains rotated by 90°. Reprinted with permission from [48], © (2019) by the American Physical Society. (**b**) Neutron scattering intensity at $\hbar\omega = 6$ meV and $T = 10$ K for $La_{2-x}Ba_xCuO_4$ with $x = 1/8$. Here, $\mathbf{Q}_{AF} = (0.5, 0.5)$; inset shows relative orientations of axes in (**a**,**b**). (**c**) Fitted dispersion and Q widths of magnetic scattering in LBCO $x = 1/8$ at 10 K along $\mathbf{Q} = (H, 0.5)$. Black line shows the hourglass dispersion often applied to such data. (**b**,**c**) Reprinted with permission from [49], ©(2007) by the American Physical Society.

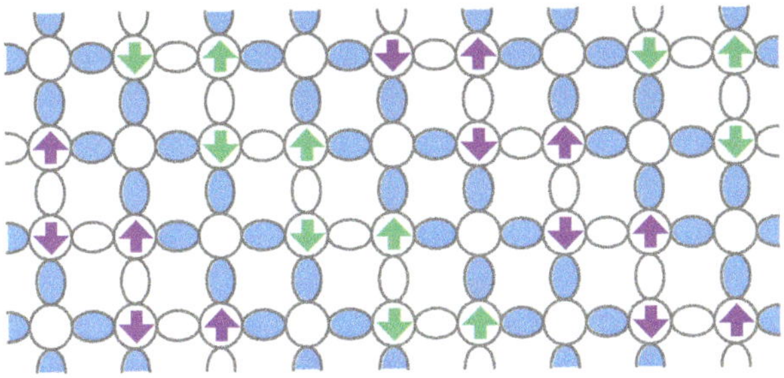

Figure 3. Diagonal stripe order as observed in $La_{2-x}Sr_xNiO_4$ with $x = 1/3$. Arrows indicate relative spin orientations on Ni sites (circles), with color change indicating antiphase domains. Blue shading indicates distribution of doped holes on O sites (ellipses).

In 214 cuprates, the stripe orientation rotates from diagonal to bond-parallel, and superconductivity appears for $x \gtrsim 0.05$ [5,10,15]. The stripe order is stabilized by coupling to lattice anisotropy, with the strongest stripe order correlated with a strong suppression of three-dimensional superconducting order at $x \approx 1/8$ [4,9]. The static spin order and the low-energy magnetic excitations correspond to the antiphase spin-stripe domains of Figure 1; an example is shown in Figure 2b. The absence of any other low-energy magnetic excitations indicates that the spin degrees of freedom on the charge stripes are gapped. The size of the gap at $\mathbf{Q}_{\mathrm{AF}}$, apparent in Figure 2c, is ~50 meV, above which commensurate AF excitations appear [50]; the effective correlation length for the high-energy excitations is only about one lattice spacing [49]. (A two-component picture of the magnetic excitations was also proposed in [51].)

We can reconcile the variations in the magnetic spectra through the model indicated schematically in Figure 4. If the charge stripes are centered on a row of bridging O atoms, then the charge stripes are effectively 2-leg spin ladders that are decoupled from the neighboring spin stripes due to frustration of the AF coupling [22]. An undoped spin ladder is a spin liquid [52], with a spin gap that can be as large as $J/2$ [23]. The hole concentration in the 2-leg ladder picture of the charge stripes is 25%. With an effective J of ~100 meV [50], the holes form pairs so as to avoid exciting the spins across the large spin gap. As illustrated in Figure 4, the spins can be viewed as forming a resonating-valence-bond (RVB) state of nearest-neighbor singlets. Theoretical analysis indicates that the singlet-triplet excitation energy is essentially the pairing scale for the doped holes, and the pairs have d-wave-like character [24,53].

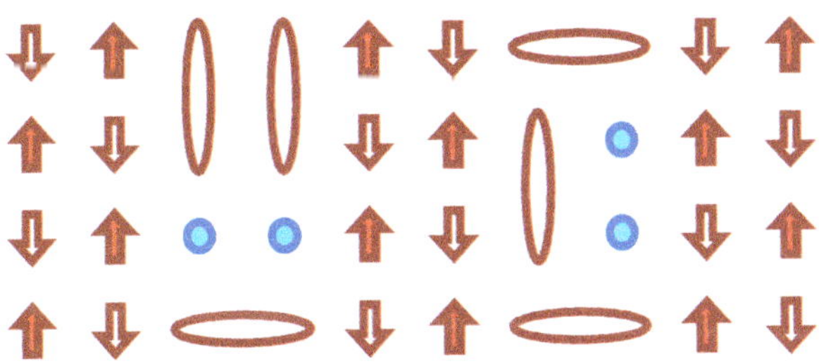

Figure 4. Cartoon of cuprate spin stripe order at $p = 1/8$, with resulting pairing correlations within the charge stripes, as proposed in [22]. Here, only Cu sites are shown. Arrows indicate ordered spins; blue circles are doped holes; ellipses are spin singlets on pairs of Cu sites.

Note that the RVB state of the 2-leg ladder is a gapped spin liquid, in contrast to the gapless quantum spin liquid of Anderson's proposed RVB for the 2D square lattice [54]. It is closer to the short-range RVB of Kivelson, Rokshar, and Sethna [55], in which a coupling to nearest-neighbor bond-length fluctuations (Peierls mechanism) stabilizes the singlet correlations. In the stripe case, the charge segregation that enables the doped ladders is stabilized by soft phonons and a lattice distortion [9,56–59].

The distinction between the spin stripes and the doped ladders breaks down for excitations above the pairing scale. Such high-energy excitations can occur anywhere in the plane, and at such energies, the holes are no longer confined to pairs within charge stripes. The strong scattering between spins and holes leads to the short correlation length at high energies.

It is recognized that superconducting and charge-density-wave correlations compete with one another in 1D [60]. Recent calculations on 2-leg ladder models suggest that superconducting correlations survive in a 1-band Hubbard model [61] but not in a 3-band Hubbard model [62]. In the experimental case of interest, we do not have individual ladders; while the spin components are decoupled by the magnetic topology, the holes in neighboring ladders interact by long-range Coulomb repulsion and possibly other effects not considered in the calculations. Furthermore, superconducting order requires Josephson coupling between the charge stripes [25]. So the important question is, what happens regarding superconductivity in experiments?

3. PDW Order

In the case of optimal stripe order, LBCO with $x = 1/8$, 2D superconducting correlations were observed to set in together with the spin-stripe order at $\sim$40 K [30,63]; 2D superconducting order was established through a Berezinskii–Kosterlitz–Thouless transition at 16 K, with 3D superconductivity developing only at $\sim$5 K. Related behavior was observed in Nd-doped LSCO with $x = 0.15$ [64], where the transition to the crystal structure that pins stripe order can be tuned with increasing Nd concentration [65]; measurements of c-axis optical conductivity demonstrated the loss of 3D superconductivity as the Nd concentration was tuned through the structural transition [31].

To explain the 2D superconductivity, a novel superconducting state was proposed: pair-density-wave order [28,29]. In the PDW state, the pair wave function oscillates from positive on one charge stripe to negative on the next, passing through zero in the spin stripes. Because the stripe order is pinned to a lattice anisotropy that rotates by 90° on passing from one layer to the next along the c axis [66], the interlayer Josephson coupling associated with PDW order should be frustrated.

The PDW order is characterized by a finite wave vector that matches that of the spin stripe order. This finite-momentum of pairs is shared with the concept of superconductivity in a strong uniform magnetic field proposed by Fulde and Farrel [67] and Larkin and Ochinnikov [68]; the difference is the absence of a net magnetic field. (Experimental evidence for a field-induced FFLO state in a layered organic superconductor was reported fairly recently [69,70].) It is also apt to note that there have been other proposals for pairing based on charge-density waves (CDWs) in cuprates. In particular, Castellani, Di Castro, Grilli, and coworkers [71–73] proposed that dynamical CDWs underlie the superconductivity of cuprates. The fluctuating CDWs would provide a pairing interaction between extended quasiparticles as in the general case of bosonic fluctuations near a quantum critical point [74]. From this perspective, static CDW order tends to compete with superconductivity.

The evidence of 2D superconductivity in LBCO, together with the PDW proposal, supported the alternative concept of intertwined order [75]. Here the idea is that the interactions that drive pairing and spin order actually work together, but benefit from spatial segregation. This approach builds on theoretical evidence that static spatial inhomogeneity can enhance pairing [76–78]. The concept of pairing within charge stripes has also had to evolve. The initial proposal for pairing in charge stripes relied on interacting with a spin gap in the neighboring spin stripes [25], which is not consistent with the presence of spin stripe order. The idea of charge stripes as doped 2-leg spin ladders resolves this problem [22].

While the proposed PDW state can explain the 2D superconductivity in LBCO, the story would be more compelling with direct evidence for PDW order in LBCO. Phase-sensitive evidence has now been reported [32]. Yang [79] predicted that one could (at least partially) restore the interlayer Josephson coupling by application of an in-plane magnetic field, and that the maximum effect would occur with the field at 45° to the in-plane Cu–O bonds. This angular dependence of the superconducting critical current density along the c axis is now confirmed by experiment [32]. Hence, PDW order coexisting with spin stripe order is experimentally verified. The conclusion that the charge stripes are the source of pairing seems unavoidable.

Further evidence of the last conclusion comes from transport measurements in a large magnetic field applied perpendicular to the planes. Such measurements on LBCO $x = 1/8$ revealed, beyond a reentrant 2D superconducting phase at a field of 20 T, an ultra-quantum metal phase with a very large sheet resistance (twice the quantum of resistance for pairs) that appears to saturate at low temperature [33]. The Hall resistance in this phase, as in the 3D and 2D superconducting phases, is zero within the error bars. (Similar results were obtained for $La_{1.7}Eu_{0.2}Sr_{0.10}CuO_4$ and $La_{1.48}Nd_{0.4}Sr_{0.12}CuO_4$ [80]). A possible interpretation of the Hall resistance at high field is that the doped holes remain paired, even with the loss of PDW phase coherence between neighboring charge stripes. Theoretically, if one takes the disorder into account, this could be a Bose metal phase [81].

Incoherent stripe correlations are also present in the normal state of LSCO [7,8]. Intriguingly, a study of shot noise in tunnel junctions involving LSCO films found evidence for pairs in the the normal state of underdoped samples [82]. That result is at least compatible with the concept of pair correlations in the charge stripes, even at $T > T_c$.

4. PDW vs. Uniform *d*-Wave Superconductivity

Dynamic topological doping, in the form of incommensurate spin excitations, is a common feature of underdoped cuprates [83]. Spectroscopically, the differences between cuprates with PDW order [84–86] and those with uniform d-wave superconductivity [87,88] are small, while charge stripes, static or dynamic, are common [1–3]. Hence, it seems quite reasonable to propose that charge stripes are the common pairing centers.

The difference between the PDW and uniform superconducting states is associated with the presence or absence of static spin-stripe order. In the PDW state, the static spin-stripe order is essential for the pair correlations to develop (anti-)phase coherence between neighboring charge stripes; purely fluctuating spin stripes oppose superconducting phase order. We should take a moment to acknowledge that it is surprising that we can have such spin order at all. There is already a large tendency toward spin fluctuations in the undoped CuO_2 planes [89], while 1D spin chains have no static order. There must be some degree of spin anisotropy present in order for the spin stripes to order. Besides being an open question, this represents a challenge for simulations using the Hubbard or t-J models, as they lack any term that would tend to induce spin order. As a consequence, attempts to identify PDW order in numerical simulations have generally been unsuccessful [90].

While the spin stripes are good for isolating the doped spin ladders that yield pairing, they stand in the way of spatially uniform superconducting order. If they can be gapped, then it should be possible to develop a uniform superconducting phase among electronic states at energies below the spin gap. Indeed, an analysis of the available experimental results on cuprate families indicates that the energy gap for incommensurate spin excitations is an upper limit for the superconducting gap associated with long-range coherence [34]. Note that the local pairing scale within the charge stripes will be larger than the coherent gap of the uniform order. This is consistent with observations by angle-resolved photoemission [91–93] and Raman scattering [94,95] of antinodal gaps that are much larger than the scale of the coherent gap [22]. On the theory side, recent density-matrix-renormalization-group calculations of the Hubbard model (on a lattice of width 4 or 6 Cu sites, with boundaries joined to form a cylinder) found that a modulation of the hopping between neighboring sites in one direction (around the circumference of the cylinder), as one might expect for charge stripes without spin order, enhances the superconducting correlations [78].

The close relationship between the PDW and uniform superconducting orders is illustrated by a study of the phase transitions in LBCO $x = 0.115$ as a function of uniaxial strain [96]. In the absence of strain, bulk susceptibility measurements suggest an onset of 2D superconducting correlations, along with spin-stripe order, near 40 K; however, the spin-stripe order is weaker than at $x = 1/8$ so that 3D superconducting order develops below $\sim$12 K [5]. Application of significant in-plane stress causes the bulk superconducting T_c to rise to 32 K, while muon-spin-rotation spectra indicate a reduction in the magnetically ordered volume fraction by more than 50%, consistent with a decrease in the volume of spin-stripe order and associated PDW order [96]. While the dominant character of the superconducting state changes under strain, the onset temperature of superconducting coherence changes relatively little.

Another connection is seen through the impact of proton irradiation on LBCO $x = 1/8$ [97]. Protons create narrow tracks of structural defects, often used as pinning centers for magnetic vortices. In LBCO, moderate proton irradiation resulted in an increase in the bulk T_c, from 4 K to 6 K, while also reducing the correlation length of the charge stripes. It is difficult to see how the structural disorder induced by the bombardment would directly enhance pairing. Instead, the induced disorder must modify the coherent coupling among the correlated pairs already present.

Lee [98] proposed that PDW order is the dominant order in cuprates and that it explains the pseudogap behavior. While the proposal is interesting, there are a number of issues with it. For one thing, the PDW order in LBCO $x = 1/8$ sets in at a temperature far below the T^* crossover temperature associated with pseudogap phenomena. While there are dynamic charge and spin stripes at higher temperatures [4,99], and there could be pairing correlations within those dynamic charge stripes, there is no evidence of coherence of pairs between neighboring charge stripes, which would be essential for a reasonable definition of PDW correlations, let alone PDW order. For another thing, the PDW order as defined in [26] is not generic to most cuprate families. For example, in $YBa_2Cu_3O_{6+x}$, the charge-density-wave order develops together with a gap in the incommensurate spin excitations [100]. As discussed above, the spin gap is compatible with the uniform d-wave superconductivity that orders at lower temperatures.

Another distinction between different cuprates concerns the magnitude of the wave vector Q_{co} for the charge order and its variation with doping. In 214 cuprates, Q_{co} grows linearly with hole density p up to $p \approx 1/8$, where it saturates at $\approx$1/8 reciprocal lattice unit (rlu) [5,101]. Common contrasting behavior is typified by $YBa_2Cu_3O_{6+x}$, where Q_{co} starts at $\sim$0.34 rlu for $p \approx 0.08$, and then decreases by about $\sim$10% with doping [100,102]. These distinct doping dependencies raise questions about the relationship between the orders in different compounds.

A new study of the doping and temperature dependence of the charge-stripe order in $La_{1.8-x}Eu_{0.2}Sr_xCuO_4$ brings new insight to this issue [103]. At low temperature, where both charge- and spin-stripe orders are observed, Q_{co} follows the behavior identified in other 214 systems. With rising temperature, however, Q_{co} tends to grow in the disordered regime, especially at smaller hole density. A physically-inspired Landau–Ginzburg model, when fitted to the temperature-dependent Q_{co} measurements, provides an extrapolation that, at $T \sim 400$ K, shows behavior of Q_{co} vs. p very similar to that found in YBCO [100, 102]. Hence, it is plausible that the charge orders found in various cuprate families have a common origin.

5. PDW Around Defects

Scanning tunneling microscopy (STM) experiments have provided evidence for a local coexistence of PDW and uniform superconducting orders around defects in the superconducting order, such as magnetic vortex cores [35,104]. The detected signature corresponds to an induced charge modulation that results from the superimposed, locally coexisting orders [26]. The main system studied by STM is $Bi_2Sr_2CaCu_2O_{8+\delta}$, which tends to have a large spin gap [105]. Local, short-range PDW correlations were also detected through spatial modulation of the superconducting gap [36].

The absence of spin-stripe order removes the conflict between PDW and uniform superconducting orders discussed above. At the same time, the PDW order detected by STM appears to be induced by local defects, which is distinct from the PDW ground state detected in LBCO, where spin-stripe order appears to be an intrinsic component. A defect such as a magnetic vortex or a Zn atom substituted for Cu causes the superconducting order to go to zero [106]; that may make PDW order energetically favored in the local environment [107].

This line of reasoning provides an interesting connection to the LBCO system. A crystal with $x = 0.095$ shows a bulk $T_c = 32$ K in zero magnetic field, along with weakened stripe order relative to $x = 1/8$ [5]. Application of a c-axis magnetic field causes an enhancement of stripe order and a decoupling of the superconducting planes [108,109], presumably due to the dominance of PDW order. The regions of uniform superconductivity should act as pinning centers for the magnetic vortices since one can obtain an energy gain from inducing PDW order there.

If Zn defects act like magnetic vortices in terms of locally favoring PDW order, then enough Zn should, like the magnetic field, cause a decoupling of superconducting planes. Indeed, this effect was confirmed in a crystal of LBCO $x = 0.095$ with 1% Zn [110]. Similar behavior was observed in $La_{2-x}Sr_xCuO_4$ with $x = 0.13$ and 1% Fe [111]. One difference between Zn defects and magnetic vortices is that Zn is known to induce pinning of spin-stripe order [112]; however, it may lead to a reduction in the spin-stripe-ordering temperature when introduced to a system that already has strong spin-stripe order [113].

6. Relations to Other Superconductors

Topological doping is important in cuprates because it establishes regions of reduced dimensionality where pairing can develop in the presence of repulsive interactions. There is a natural connection with systems in which the lattice is formed from a coupling of lower-dimensional components. One example is alkali-doped C_{60} [114], where the dopants provide the charge carriers, while the the C_{60} molecules provide the interactions. Chakravarty and Kivelson [115] proposed a model in which electrons could minimize repulsive interactions by hopping onto C_{60} molecules in pairs. In fact, they made a direct comparison to pairing in a doped 2-leg spin ladder. Another obvious parallel is with organic superconductors [116], where superconductivity tends to occur in proximity to spin-density-wave order [117,118].

The situation is different if we compare with other layered superconductors. In electron-doped cuprates, such as $Nd_{2-x}Ce_xCuO_4$, the carriers and the Cu moments do not spatially segregate. As a result, commensurate antiferromagnetic order survives to a higher carrier concentration, and superconductivity appears only when static order disappears [119]. There is a good deal of commensurate inelastic magnetic spectral weight at low energy. If this were a good thing for superconductivity, then one might expect to obtain a very high T_c; instead, the highest T_c is lower than that in essentially all hole-doped families of cuprate superconductors. Neutron scattering studies on electron-doped superconductors show that the low-energy antiferromagnetic excitations become gapped on an energy scale comparable to the superconducting gap [120–122].

Funding: Work at Brookhaven is supported by the Office of Basic Energy Sciences, Materials Sciences and Engineering Division, U.S. Department of Energy (DOE) under Contract No. DE-SC0012704.

Institutional Review Board Statement: Not applicable.

Informed Consent Statement: Not applicable.

Conflicts of Interest: The author declares no conflict of interest.

References

1. Frano, A.; Blanco-Canosa, S.; Keimer, B.; Birgeneau, R.J. Charge ordering in superconducting copper oxides. *J. Phys. Condens. Matter* **2020**, *32*, 374005. [CrossRef] [PubMed]
2. Comin, R.; Damascelli, A. Resonant X-ray Scattering Studies of Charge Order in Cuprates. *Annu. Rev. Condens. Matter Phys.* **2016**, *7*, 369–405. [CrossRef]
3. Uchida, S.i. Ubiquitous Charge Order Correlations in High-Temperature Superconducting Cuprates. *J. Phys. Soc. Jpn.* **2021**, *90*, 111001. [CrossRef]
4. Fujita, M.; Goka, H.; Yamada, K.; Tranquada, J.M.; Regnault, L.P. Stripe order, depinning, and fluctuations in $La_{1.875}Ba_{0.125}CuO_4$ and $La_{1.875}Ba_{0.075}Sr_{0.050}CuO_4$. *Phys. Rev. B* **2004**, *70*, 104517. [CrossRef]
5. Hücker, M.; van Zimmermann, M.; Gu, G.D.; Xu, Z.J.; Wen, J.S.; Xu, G.; Kang, H.J.; Zheludev, A.; Tranquada, J.M. Stripe order in superconducting $La_{2-x}Ba_xCuO_4$ ($0.095 \leq x \leq 0.155$). *Phys. Rev. B* **2011**, *83*, 104506. [CrossRef]
6. Fujita, M.; Hiraka, H.; Matsuda, M.; Matsuura, M.; Tranquada, J.; Wakimoto, S.; Xu, G.; Yamada, K. Progress in Neutron Scattering Studies of Spin Excitations in High-T_c Cuprates. *J. Phys. Soc. Jpn.* **2012**, *81*, 011007. [CrossRef]
7. Wen, J.J.; Huang, H.; Lee, S.J.; Jang, H.; Knight, J.; Lee, Y.S.; Fujita, M.; Suzuki, K.M.; Asano, S.; Kivelson, S.A.; et al. Observation of two types of charge-density-wave orders in superconducting $La_{2-x}Sr_xCuO_4$. *Nat. Commun.* **2019**, *10*, 3269. [CrossRef] [PubMed]
8. Miao, H.; Fabbris, G.; Koch, R.J.; Mazzone, D.G.; Nelson, C.S.; Acevedo-Esteves, R.; Gu, G.D.; Li, Y.; Yilimaz, T.; Kaznatcheev, K.; et al. Charge density waves in cuprate superconductors beyond the critical doping. *NPJ Quantum Mater.* **2021**, *6*, 31. [CrossRef]
9. Tranquada, J.M.; Sternlieb, B.J.; Axe, J.D.; Nakamura, Y.; Uchida, S. Evidence for stripe correlations of spins and holes in copper oxide superconductors. *Nature* **1995**, *375*, 561–563. [CrossRef]
10. Ma, Q.; Rule, K.C.; Cronkwright, Z.W.; Dragomir, M.; Mitchell, G.; Smith, E.M.; Chi, S.; Kolesnikov, A.I.; Stone, M.B.; Gaulin, B.D. Parallel spin stripes and their coexistence with superconducting ground states at optimal and high doping in $La_{1.6-x}Nd_{0.4}Sr_xCuO_4$. *Phys. Rev. Res.* **2021**, *3*, 023151. [CrossRef]
11. Kivelson, S.; Emery, V. Topological doping of correlated insulators. *Synth. Met.* **1996**, *80*, 151–158. [CrossRef]
12. Hasan, M.Z.; Kane, C.L. Colloquium: Topological Insulators. *Rev. Mod. Phys.* **2010**, *82*, 3045–3067. [CrossRef]
13. Qi, X.L.; Zhang, S.C. Topological insulators and superconductors. *Rev. Mod. Phys.* **2011**, *83*, 1057–1110. [CrossRef]
14. Tranquada, J.M. Spins, stripes, and superconductivity in hole-doped cuprates. *AIP Conf. Proc.* **2013**, *1550*, 114–187. [CrossRef]
15. Fujita, M.; Yamada, K.; Hiraka, H.; Gehring, P.M.; Lee, S.H.; Wakimoto, S.; Shirane, G. Static magnetic correlations near the insulating-superconducting phase boundary in $La_{2-x}Sr_xCuO_4$. *Phys. Rev. B* **2002**, *65*, 064505. [CrossRef]
16. Pryadko, L.P.; Kivelson, S.A.; Emery, V.J.; Bazaliy, Y.B.; Demler, E.A. Topological doping and the stability of stripe phases. *Phys. Rev. B* **1999**, *60*, 7541–7557. [CrossRef]
17. Zaanen, J.; Gunnarsson, O. Charged magnetic domain lines and the magnetism of high-T_c oxides. *Phys. Rev. B* **1989**, *40*, 7391. [CrossRef]
18. Löw, U.; Emery, V.J.; Fabricius, K.; Kivelson, S.A. Study of an Ising model with competing long- and short-range interactions. *Phys. Rev. Lett.* **1994**, *72*, 1918–1921. [CrossRef]
19. White, S.R.; Scalapino, D.J. Density Matrix Renormalization Group Study of the Striped Phase in the 2D $t-J$ Model. *Phys. Rev. Lett.* **1998**, *80*, 1272–1275. [CrossRef]
20. Zheng, B.X.; Chung, C.M.; Corboz, P.; Ehlers, G.; Qin, M.P.; Noack, R.M.; Shi, H.; White, S.R.; Zhang, S.; Chan, G.K.L. Stripe order in the underdoped region of the two-dimensional Hubbard model. *Science* **2017**, *358*, 1155–1160. [CrossRef] [PubMed]
21. Huang, E.W.; Mendl, C.B.; Jiang, H.C.; Moritz, B.; Devereaux, T.P. Stripe order from the perspective of the Hubbard model. *NPJ Quantum Mater.* **2018**, *3*, 22. [CrossRef]
22. Tranquada, J.M. Cuprate superconductors as viewed through a striped lens. *Adv. Phys.* **2020**, *69*, 437–509. [CrossRef]
23. Dagotto, E.; Rice, T.M. Surprises on the Way from One- to Two-Dimensional Quantum Magnets: The Ladder Materials. *Science* **1996**, *271*, 618–623. [CrossRef]
24. Tsvelik, A.M. Field theory for a fermionic ladder with generic intrachain interactions. *Phys. Rev. B* **2011**, *83*, 104405. [CrossRef]
25. Emery, V.J.; Kivelson, S.A.; Zachar, O. Spin-gap proximity effect mechanism of high-temperature superconductivity. *Phys. Rev. B* **1997**, *56*, 6120–6147. [CrossRef]
26. Berg, E.; Fradkin, E.; Kivelson, S.A.; Tranquada, J.M. Striped superconductors: how spin, charge and superconducting orders intertwine in the cuprates. *New J. Phys.* **2009**, *11*, 115004. [CrossRef]
27. Agterberg, D.F.; Davis, J.S.; Edkins, S.D.; Fradkin, E.; Van Harlingen, D.J.; Kivelson, S.A.; Lee, P.A.; Radzihovsky, L.; Tranquada, J.M.; Wang, Y. The Physics of Pair-Density Waves: Cuprate Superconductors and Beyond. *Annu. Rev. Condens. Matter Phys.* **2020**, *11*, 231–270. [CrossRef]
28. Himeda, A.; Kato, T.; Ogata, M. Stripe States with Spatially Oscillating *d*-Wave Superconductivity in the Two-Dimensional $t-t'-J$ Model. *Phys. Rev. Lett.* **2002**, *88*, 117001. [CrossRef]
29. Berg, E.; Fradkin, E.; Kim, E.A.; Kivelson, S.A.; Oganesyan, V.; Tranquada, J.M.; Zhang, S.C. Dynamical Layer Decoupling in a Stripe-Ordered High-T_c Superconductor. *Phys. Rev. Lett.* **2007**, *99*, 127003. [CrossRef] [PubMed]
30. Li, Q.; Hücker, M.; Gu, G.D.; Tsvelik, A.M.; Tranquada, J.M. Two-Dimensional Superconducting Fluctuations in Stripe-Ordered $La_{1.875}Ba_{0.125}CuO_4$. *Phys. Rev. Lett.* **2007**, *99*, 067001. [CrossRef] [PubMed]
31. Tajima, S.; Noda, T.; Eisaki, H.; Uchida, S. *c*-Axis Optical Response in the Static Stripe Ordered Phase of the Cuprates. *Phys. Rev. Lett.* **2001**, *86*, 500–503. [CrossRef] [PubMed]

32. Lozano, P.M.; Ren, T.; Gu, G.D.; Tsvelik, A.M.; Tranquada, J.M.; Li, Q. Phase-sensitive evidence of pair-density-wave order in a cuprate. *arXiv* **2021**, arXiv:2110.05513.
33. Li, Y.; Terzic, J.; Baity, P.G.; Popović, D.; Gu, G.D.; Li, Q.; Tsvelik, A.M.; Tranquada, J.M. Tuning from failed superconductor to failed insulator with magnetic field. *Sci. Adv.* **2019**, *5*, eaav7686. [CrossRef] [PubMed]
34. Li, Y.; Zhong, R.; Stone, M.B.; Kolesnikov, A.I.; Gu, G.D.; Zaliznyak, I.A.; Tranquada, J.M. Low-energy antiferromagnetic spin fluctuations limit the coherent superconducting gap in cuprates. *Phys. Rev. B* **2018**, *98*, 224508. [CrossRef]
35. Edkins, S.D.; Kostin, A.; Fujita, K.; Mackenzie, A.P.; Eisaki, H.; Uchida, S.; Sachdev, S.; Lawler, M.J.; Kim, E.A.; Séamus Davis, J.C.; et al. Magnetic field–induced pair density wave state in the cuprate vortex halo. *Science* **2019**, *364*, 976–980. [CrossRef] [PubMed]
36. Du, Z.; Li, H.; Joo, S.H.; Donoway, E.P.; Lee, J.; Davis, J.C.S.; Gu, G.; Johnson, P.D.; Fujita, K. Imaging the energy gap modulations of the cuprate pair-density-wave state. *Nature* **2020**, *580*, 65–70. [CrossRef]
37. Chen, C.T.; Sette, F.; Ma, Y.; Hybertsen, M.S.; Stechel, E.B.; Foulkes, W.M.C.; Schulter, M.; Cheong, S.W.; Cooper, A.S.; Rupp, L.W.; et al. Electronic states in $La_{2-x}Sr_xCuO_{4+\delta}$ probed by soft-X-ray absorption. *Phys. Rev. Lett.* **1991**, *66*, 104–107. [CrossRef]
38. Emery, V.J.; Reiter, G. Mechanism for high-temperature superconductivity. *Phys. Rev. B* **1988**, *38*, 4547–4556. [CrossRef]
39. Matsuda, M.; Fujita, M.; Yamada, K.; Birgeneau, R.J.; Endoh, Y.; Shirane, G. Electronic phase separation in lightly doped $La_{2-x}Sr_xCuO_4$. *Phys. Rev. B* **2002**, *65*, 134515. [CrossRef]
40. Vajk, O.P.; Mang, P.K.; Greven, M.; Gehring, P.M.; Lynn, J.W. Quantum Impurities in the Two-Dimensional Spin One-Half Heisenberg Antiferromagnet. *Science* **2002**, *295*, 1691–1695. [CrossRef]
41. Tranquada, J.M.; Buttrey, D.J.; Sachan, V.; Lorenzo, J.E. Simultaneous Ordering of Holes and Spins in $La_2NiO_{4.125}$. *Phys. Rev. Lett.* **1994**, *73*, 1003–1006. [CrossRef] [PubMed]
42. Yoshizawa, H.; Kakeshita, T.; Kajimoto, R.; Tanabe, T.; Katsufuji, T.; Tokura, Y. Stripe order at low temperatures in $La_{2-x}Sr_xNiO_4$ with $0.289 \lesssim x \lesssim 0.5$. *Phys. Rev. B* **2000**, *61*, R854–R857. [CrossRef]
43. Ulbrich, H.; Braden, M. Neutron scattering studies on stripe phases in non-cuprate materials. *Phys. C Supercond.* **2012**, *481*, 31–45. [CrossRef]
44. Lee, S.H.; Cheong, S.W. Melting of Quasi-Two-Dimensional Charge Stripes in $La_{5/3}Sr_{1/3}NiO_4$. *Phys. Rev. Lett.* **1997**, *79*, 2514–2517. [CrossRef]
45. Boothroyd, A.T.; Prabhakaran, D.; Freeman, P.G.; Lister, S.J.S.; Enderle, M.; Hiess, A.; Kulda, J. Spin dynamics in stripe-ordered $La_{5/3}Sr_{1/3}NiO_4$. *Phys. Rev. B* **2003**, *67*, 100407(R). [CrossRef]
46. Woo, H.; Boothroyd, A.T.; Nakajima, K.; Perring, T.G.; Frost, C.D.; Freeman, P.G.; Prabhakaran, D.; Yamada, K.; Tranquada, J.M. Mapping spin-wave dispersions in stripe-ordered $La_{2-x}Sr_xNiO_4$ ($x = 0.275, 0.333$). *Phys. Rev. B* **2005**, *72*, 064437. [CrossRef]
47. Boothroyd, A.T.; Freeman, P.G.; Prabhakaran, D.; Hiess, A.; Enderle, M.; Kulda, J.; Altorfer, F. Spin Correlations among the Charge Carriers in an Ordered Stripe Phase. *Phys. Rev. Lett.* **2003**, *91*, 257201. [CrossRef]
48. Merritt, A.M.; Reznik, D.; Garlea, V.O.; Gu, G.D.; Tranquada, J.M. Nature and impact of stripe freezing in $La_{1.67}Sr_{0.33}NiO_4$. *Phys. Rev. B* **2019**, *100*, 195122. [CrossRef]
49. Xu, G.; Tranquada, J.M.; Perring, T.G.; Gu, G.D.; Fujita, M.; Yamada, K. High-energy magnetic excitations from dynamic stripes in $La_{1.875}Ba_{0.125}CuO_4$. *Phys. Rev. B* **2007**, *76*, 014508. [CrossRef]
50. Tranquada, J.M.; Woo, H.; Perring, T.G.; Goka, H.; Gu, G.D.; Xu, G.; Fujita, M.; Yamada, K. Quantum magnetic excitations from stripes in copper oxide superconductors. *Nature* **2004**, *429*, 534–538. [CrossRef] [PubMed]
51. Sato, K.; Ikeuchi, K.; Kajimoto, R.; Wakimoto, S.; Arai, M.; Fujita, M. Coexistence of Two Components in Magnetic Excitations of $La_{2-x}Sr_xCuO_4$ ($x = 0.10$ and 0.16). *J. Phys. Soc. Jpn.* **2020**, *89*, 114703. [CrossRef]
52. Barnes, T.; Dagotto, E.; Riera, J.; Swanson, E.S. Excitation spectrum of Heisenberg spin ladders. *Phys. Rev. B* **1993**, *47*, 3196–3203. [CrossRef]
53. Poilblanc, D.; Scalapino, D.J.; Capponi, S. Superconducting Gap for a Two-Leg $t-J$ Ladder. *Phys. Rev. Lett.* **2003**, *91*, 137203. [CrossRef] [PubMed]
54. Anderson, P.W. The Resonating Valence Bond State in La_2CuO_4 and Superconductivity. *Science* **1987**, *235*, 1196–1198. [CrossRef]
55. Kivelson, S.A.; Rokhsar, D.S.; Sethna, J.P. Topology of the resonating valence-bond state: Solitons and high-T_c superconductivity. *Phys. Rev. B* **1987**, *35*, 8865(R). [CrossRef] [PubMed]
56. Reznik, D.; Pintschovius, L.; Ito, M.; Iikubo, S.; Sato, M.; Goka, H.; Fujita, M.; Yamada, K.; Gu, G.D.; Tranquada, J.M. Electron–phonon coupling reflecting dynamic charge inhomogeneity in copper-oxide superconductors. *Nature* **2006**, *440*, 1170. [CrossRef]
57. Miao, H.; Ishikawa, D.; Heid, R.; Le Tacon, M.; Fabbris, G.; Meyers, D.; Gu, G.D.; Baron, A.Q.R.; Dean, M.P.M. Incommensurate Phonon Anomaly and the Nature of Charge Density Waves in Cuprates. *Phys. Rev. X* **2018**, *8*, 011008. [CrossRef]
58. Peng, Y.Y.; Husain, A.A.; Mitrano, M.; Sun, S.X.L.; Johnson, T.A.; Zakrzewski, A.V.; MacDougall, G.J.; Barbour, A.; Jarrige, I.; Bisogni, V.; et al. Enhanced Electron-Phonon Coupling for Charge-Density-Wave Formation in $La_{1.8-x}Eu_{0.2}Sr_xCuO_{4+\delta}$. *Phys. Rev. Lett.* **2020**, *125*, 097002. [CrossRef] [PubMed]
59. Wang, Q.; von Arx, K.; Horio, M.; Mukkattukavil, D.J.; Küspert, J.; Sassa, Y.; Schmitt, T.; Nag, A.; Pyon, S.; Takayama, T.; et al. Charge order lock-in by electron-phonon coupling in $La_{1.675}Eu_{0.2}Sr_{0.125}CuO_4$. *Sci. Adv.* **2021**, *7*, eabg7394. [CrossRef]
60. Kivelson, S.A.; Fradkin, E.; Emery, V.J. Electronic liquid-crystal phases of a doped Mott insulator. *Nature* **1998**, *393*, 550–553. [CrossRef]
61. Dolfi, M.; Bauer, B.; Keller, S.; Troyer, M. Pair correlations in doped Hubbard ladders. *Phys. Rev. B* **2015**, *92*, 195139. [CrossRef]

62. Song, J.P.; Mazumdar, S.; Clay, R.T. Absence of Luther-Emery superconducting phase in the three-band model for cuprate ladders. *Phys. Rev. B* **2021**, *104*, 104504. [CrossRef]
63. Tranquada, J.M.; Gu, G.D.; Hücker, M.; Jie, Q.; Kang, H.J.; Klingeler, R.; Li, Q.; Tristan, N.; Wen, J.S.; Xu, G.Y.; et al. Evidence for unusual superconducting correlations coexisting with stripe order in $La_{1.875}Ba_{0.125}CuO_4$. *Phys. Rev. B* **2008**, *78*, 174529. [CrossRef]
64. Ding, J.F.; Xiang, X.Q.; Zhang, Y.Q.; Liu, H.; Li, X.G. Two-dimensional superconductivity in stripe-ordered $La_{1.6-x}Nd_{0.4}Sr_xCuO_4$ single crystals. *Phys. Rev. B* **2008**, *77*, 214524. [CrossRef]
65. Büchner, B.; Breuer, M.; Freimuth, A.; Kampf, A.P. Critical Buckling for the Disappearance of Superconductivity in Rare-Earth-Doped $La_{2-x}Sr_xCuO_4$. *Phys. Rev. Lett.* **1994**, *73*, 1841–1844. [CrossRef] [PubMed]
66. Axe, J.D.; Crawford, M.K. Structural instabilities in lanthanum cuprate superconductors. *J. Low Temp. Phys.* **1994**, *95*, 271–284. [CrossRef]
67. Fulde, P.; Ferrell, R.A. Superconductivity in a Strong Spin-Exchange Field. *Phys. Rev.* **1964**, *135*, A550–A563. [CrossRef]
68. Larkin, A.I.; Ovchinnikov, Y.N. Nonuniform State of Superconductors. *Zh. Eksp. Teor. Fiz.* **1964**, *47*, 1136.
69. Koutroulakis, G.; Kühne, H.; Schlueter, J.A.; Wosnitza, J.; Brown, S.E. Microscopic Study of the Fulde-Ferrell-Larkin-Ovchinnikov State in an All-Organic Superconductor. *Phys. Rev. Lett.* **2016**, *116*, 067003. [CrossRef]
70. Sugiura, S.; Isono, T.; Terashima, T.; Yasuzuka, S.; Schlueter, J.A.; Uji, S. Fulde–Ferrell–Larkin–Ovchinnikov and vortex phases in a layered organic superconductor. *NPJ Quantum Mater.* **2019**, *4*, 7. [CrossRef]
71. Castellani, C.; Di Castro, C.; Grilli, M. Singular Quasiparticle Scattering in the Proximity of Charge Instabilities. *Phys. Rev. Lett.* **1995**, *75*, 4650–4653. [CrossRef] [PubMed]
72. Grilli, M.; Seibold, G.; Di Ciolo, A.; Lorenzana, J. Fermi surface dichotomy in systems with fluctuating order. *Phys. Rev. B* **2009**, *79*, 125111. [CrossRef]
73. Caprara, S.; Di Castro, C.; Seibold, G.; Grilli, M. Dynamical charge density waves rule the phase diagram of cuprates. *Phys. Rev. B* **2017**, *95*, 224511. [CrossRef]
74. Wang, Y.; Abanov, A.; Altshuler, B.L.; Yuzbashyan, E.A.; Chubukov, A.V. Superconductivity near a Quantum-Critical Point: The Special Role of the First Matsubara Frequency. *Phys. Rev. Lett.* **2016**, *117*, 157001. [CrossRef]
75. Fradkin, E.; Kivelson, S.A.; Tranquada, J.M. Colloquium: Theory of intertwined orders in high temperature superconductors. *Rev. Mod. Phys.* **2015**, *87*, 457–482. [CrossRef]
76. Carlson, E.W.; Emery, V.J.; Kivelson, S.A.; Orgad, D. Concepts in High Temperature Superconductivity. In *The Physics of Superconductors Vol II: Superconductivity in Nanostructures, High-T_c and Novel Superconductors, Organic Superconducors*; Bennemann, K.H.; Ketterson, J.B., Eds.; Springer: Berlin, Germany, 2004.
77. Tsai, W.F.; Yao, H.; Läuchli, A.; Kivelson, S.A. Optimal inhomogeneity for superconductivity: Finite-size studies. *Phys. Rev. B* **2008**, *77*, 214502. [CrossRef]
78. Jiang, H.C.; Kivelson, S.A. Stripe order enhanced superconductivity in the Hubbard model. *arXiv* **2021**, arXiv:2105.07048.
79. Yang, K. Detection of Striped Superconductors Using Magnetic Field Modulated Josephson Effect. *J. Supercond. Nov. Magn.* **2013**, *26*, 2741–2742. [CrossRef]
80. Shi, Z.; Baity, P.G.; Terzic, J.; Pokharel, B.K.; Sasagawa, T.; Popović, D. Magnetic field reveals vanishing Hall response in the normal state of stripe-ordered cuprates. *Nat. Commun.* **2021**, *12*, 3724. [CrossRef]
81. Ren, T.; Tsvelik, A.M. How magnetic field can transform a superconductor into a Bose metal. *New J. Phys.* **2020**, *22*, 103021. [CrossRef]
82. Zhou, P.; Chen, L.; Liu, Y.; Sochnikov, I.; Bollinger, A.T.; Han, M.G.; Zhu, Y.; He, X.; Božović, I.; Natelson, D. Electron pairing in the pseudogap state revealed by shot noise in copper oxide junctions. *Nature* **2019**, *572*, 493–496. [CrossRef] [PubMed]
83. Enoki, M.; Fujita, M.; Nishizaki, T.; Iikubo, S.; Singh, D.K.; Chang, S.; Tranquada, J.M.; Yamada, K. Spin-Stripe Density Varies Linearly With the Hole Content in Single-Layer $Bi_{2+x}Sr_{2-x}CuO_{6+y}$ Cuprate Superconductors. *Phys. Rev. Lett.* **2013**, *110*, 017004. [CrossRef]
84. Valla, T.; Federov, A.V.; Lee, J.; Davis, J.C.; Gu, G.D. The Ground State of the Pseudogap in Cuprate Superconductors. *Science* **2006**, *314*, 1914. [CrossRef]
85. He, R.H.; Tanaka, K.; Mo, S.K.; Sasagawa, T.; Fujita, M.; Adachi, T.; Mannella, N.; Yamada, K.; Koike, Y.; Hussain, Z.; et al. Energy gaps in the failed high-T_c superconductor $La_{1.875}Ba_{0.125}CuO_4$. *Nat. Phys.* **2009**, *5*, 119–123. [CrossRef]
86. Homes, C.C.; Hücker, M.; Li, Q.; Xu, Z.J.; Wen, J.S.; Gu, G.D.; Tranquada, J.M. Determination of the optical properties of $La_{2-x}Ba_xCuO_4$ for several dopings, including the anomalous $x = \frac{1}{8}$ phase. *Phys. Rev. B* **2012**, *85*, 134510. [CrossRef]
87. Damascelli, A.; Hussain, Z.; Shen, Z.X. Angle-resolved photoemission studies of the cuprate superconductors. *Rev. Mod. Phys.* **2003**, *75*, 473–541. [CrossRef]
88. Basov, D.N.; Timusk, T. Electrodynamics of high-T_c superconductors. *Rev. Mod. Phys.* **2005**, *77*, 721–779. [CrossRef]
89. Chakravarty, S.; Halperin, B.I.; Nelson, D.R. Low-temperature behavior of two-dimensional quantum antiferromagnets. *Phys. Rev. Lett.* **1988**, *60*, 1057–1060. [CrossRef] [PubMed]
90. White, S.R.; Scalapino, D.J. Doping asymmetry and striping in a three-orbital CuO_2 Hubbard model. *Phys. Rev. B* **2015**, *92*, 205112. [CrossRef]
91. Lee, W.S.; Vishik, I.M.; Tanaka, K.; Lu, D.H.; Sasagawa, T.; Nagaosa, N.; Devereaux, T.P.; Hussain, Z.; Shen, Z.X. Abrupt onset of a second energy gap at the superconducting transition of underdoped Bi2212. *Nature* **2007**, *450*, 81–84. [CrossRef] [PubMed]

92. Yoshida, M.; Ishii, K.; Naka, M.; Ishihara, S.; Jarrige, I.; Ikeuchi, K.; Murakami, Y.; Kudo, K.; Koike, Y.; Nagata, T.; et al. Observation of momentum-resolved charge fluctuations proximate to the charge-order phase using resonant inelastic X-ray scattering. *Sci. Rep.* **2016**, *6*, 23611. [CrossRef] [PubMed]
93. Drozdov, I.K.; Pletikosić, I.; Kim, C.K.; Fujita, K.; Gu, G.D.; Davis, J.C.S.; Johnson, P.D.; Božović, I.; Valla, T. Phase diagram of $Bi_2Sr_2CaCu_2O_{8+\delta}$ revisited. *Nat. Commun.* **2018**, *9*, 5210. [CrossRef]
94. Munnikes, N.; Muschler, B.; Venturini, F.; Tassini, L.; Prestel, W.; Ono, S.; Ando, Y.; Peets, D.C.; Hardy, W.N.; Liang, R.; et al. Pair breaking versus symmetry breaking: Origin of the Raman modes in superconducting cuprates. *Phys. Rev. B* **2011**, *84*, 144523. [CrossRef]
95. Sacuto, A.; Gallais, Y.; Cazayous, M.; Méasson, M.A.; Gu, G.D.; Colson, D. New insights into the phase diagram of the copper oxide superconductors from electronic Raman scattering. *Rep. Prog. Phys.* **2013**, *76*, 022502. [CrossRef]
96. Guguchia, Z.; Das, D.; Wang, C.N.; Adachi, T.; Kitajima, N.; Elender, M.; Brückner, F.; Ghosh, S.; Grinenko, V.; Shiroka, T.; et al. Using Uniaxial Stress to Probe the Relationship between Competing Superconducting States in a Cuprate with Spin-stripe Order. *Phys. Rev. Lett.* **2020**, *125*, 097005. [CrossRef] [PubMed]
97. Leroux, M.; Mishra, V.; Ruff, J.P.C.; Claus, H.; Smylie, M.P.; Opagiste, C.; Rodière, P.; Kayani, A.; Gu, G.D.; Tranquada, J.M.; et al. Disorder raises the critical temperature of a cuprate superconductor. *Proc. Natl. Acad. Sci. USA* **2019**, *116*, 10691–10697. [CrossRef]
98. Lee, P.A. Amperean Pairing and the Pseudogap Phase of Cuprate Superconductors. *Phys. Rev. X* **2014**, *4*, 031017. [CrossRef]
99. Miao, H.; Lorenzana, J.; Seibold, G.; Peng, Y.Y.; Amorese, A.; Yakhou-Harris, F.; Kummer, K.; Brookes, N.B.; Konik, R.M.; Thampy, V.; et al. High-temperature charge density wave correlations in $La_{1.875}Ba_{0.125}CuO_4$ without spin-charge locking. *Proc. Natl. Acad. Sci. USA* **2017**, *114*, 12430–12435. [CrossRef] [PubMed]
100. Hücker, M.; Christensen, N.B.; Holmes, A.T.; Blackburn, E.; Forgan, E.M.; Liang, R.; Bonn, D.A.; Hardy, W.N.; Gutowski, O.; Zimmermann, M.V.; et al. Competing charge, spin, and superconducting orders in underdoped $YBa_2Cu_3O_y$. *Phys. Rev. B* **2014**, *90*, 054514. [CrossRef]
101. Birgeneau, R.J.; Stock, C.; Tranquada, J.M.; Yamada, K. Magnetic Neutron Scattering in Hole-Doped Cuprate Superconductors. *J. Phys. Soc. Jpn.* **2006**, *75*, 111003. [CrossRef]
102. Blanco-Canosa, S.; Frano, A.; Schierle, E.; Porras, J.; Loew, T.; Minola, M.; Bluschke, M.; Weschke, E.; Keimer, B.; Le Tacon, M. Resonant X-ray scattering study of charge-density wave correlations in $YBa_2Cu_3O_{6+x}$. *Phys. Rev. B* **2014**, *90*, 054513. [CrossRef]
103. Lee, S.; Huang, E.W.; Johnson, T.A.; Guo, X.; Husain, A.A.; Mitrano, M.; Lu, K.; Zakrzewski, A.V.; de la Peña, G.; Peng, Y.; et al. Generic character of charge and spin density waves in superconducting cuprates. *arXiv* **2021**, arXiv:2110.13991.
104. Hoffman, J.E.; Hudson, E.W.; Lang, K.M.; Madhavan, V.; Eisaki, H.; Uchida, S.; Davis, J.C. A Four Unit Cell Periodic Pattern of Quasi-Particle States Surrounding Vortex Cores in $Bi_2Sr_2CaCu_2O_{8+\delta}$. *Science* **2002**, *295*, 466–469. [CrossRef] [PubMed]
105. Xu, G.; Gu, G.D.; Hücker, M.; Fauque, B.; Perring, T.G.; Regnault, L.P.; Tranquada, J.M. Testing the itinerancy of spin dynamics in superconducting $Bi_2Sr_2CaCu_2O_{8+\delta}$. *Nat. Phys.* **2009**, *5*, 642–646. [CrossRef]
106. Nachumi, B.; Keren, A.; Kojima, K.; Larkin, M.; Luke, G.M.; Merrin, J.; Tchernyshöv, O.; Uemura, Y.J.; Ichikawa, N.; Goto, M.; et al. Muon Spin Relaxation Studies of Zn-Substitution Effects in High-T_c Cuprate Superconductors. *Phys. Rev. Lett.* **1996**, *77*, 5421–5424. [CrossRef] [PubMed]
107. Tranquada, J.M.; Dean, M.P.M.; Li, Q. Superconductivity from Charge Order in Cuprates. *J. Phys. Soc. Jpn.* **2021**, *90*, 111002. [CrossRef]
108. Wen, J.; Jie, Q.; Li, Q.; Hücker, M.V.; Zimmermann, M.; Han, S.J.; Xu, Z.; Singh, D.K.; Konik, R.M.; Zhang, L.; et al. Uniaxial linear resistivity of superconducting $La_{1.905}Ba_{0.095}CuO_4$ induced by an external magnetic field. *Phys. Rev. B* **2012**, *85*, 134513. [CrossRef]
109. Stegen, Z.; Han, S.J.; Wu, J.; Pramanik, A.K.; Hücker, M.; Gu, G.; Li, Q.; Park, J.H.; Boebinger, G.S.; Tranquada, J.M. Evolution of superconducting correlations within magnetic-field-decoupled $La_{2-x}Ba_xCuO_4$ ($x = 0.095$). *Phys. Rev. B* **2013**, *87*, 064509. [CrossRef]
110. Lozano, P.M.; Gu, G.D.; Tranquada, J.M.; Li, Q. Experimental evidence that zinc impurities pin pair-density-wave order in $La_{2-x}Ba_xCuO_4$. *Phys. Rev. B* **2021**, *103*, L020502. [CrossRef]
111. Huang, H.; Lee, S.J.; Ikeda, Y.; Taniguchi, T.; Takahama, M.; Kao, C.C.; Fujita, M.; Lee, J.S. Two-Dimensional Superconducting Fluctuations Associated with Charge-Density-Wave Stripes in $La_{1.87}Sr_{0.13}Cu_{0.99}Fe_{0.01}O_4$. *Phys. Rev. Lett.* **2021**, *126*, 167001. [CrossRef]
112. Hirota, K. Neutron scattering studies of Zn-doped $La_{2-x}Sr_xCuO_4$. *Physica C* **2001**, *357–360*, 61–68. [CrossRef]
113. Guguchia, Z.; Roessli, B.; Khasanov, R.; Amato, A.; Pomjakushina, E.; Conder, K.; Uemura, Y.J.; Tranquada, J.M.; Keller, H.; Shengelaya, A. Complementary Response of Static Spin-Stripe Order and Superconductivity to Nonmagnetic Impurities in Cuprates. *Phys. Rev. Lett.* **2017**, *119*, 087002. [CrossRef]
114. Ganin, A.Y.; Takabayashi, Y.; Jeglič, P.; Arčon, D.; Potočnik, A.; Baker, P.J.; Ohishi, Y.; McDonald, M.T.; Tzirakis, M.D.; McLennan, A.; et al. Polymorphism control of superconductivity and magnetism in Cs_3C_{60} close to the Mott transition. *Nature* **2010**, *466*, 221–225. [CrossRef]
115. Chakravarty, S.; Kivelson, S.A. Electronic mechanism of superconductivity in the cuprates, C_{60}, and polyacenes. *Phys. Rev. B* **2001**, *64*, 064511. [CrossRef]
116. Ardavan, A.; Brown, S.; Kagoshima, S.; Kanoda, K.; Kuroki, K.; Mori, H.; Ogata, M.; Uji, S.; Wosnitza, J. Recent Topics of Organic Superconductors. *J. Phys. Soc. Jpn.* **2012**, *81*, 011004. [CrossRef]

117. Kang, N.; Salameh, B.; Auban-Senzier, P.; Jérome, D.; Pasquier, C.R.; Brazovskii, S. Domain walls at the spin-density-wave endpoint of the organic superconductor $(TMTSF)_2PF_6$ under pressure. *Phys. Rev. B* **2010**, *81*, 100509. [CrossRef]
118. Kawasugi, Y.; Seki, K.; Tajima, S.; Pu, J.; Takenobu, T.; Yunoki, S.; Yamamoto, H.M.; Kato, R. Two-dimensional ground-state mapping of a Mott-Hubbard system in a flexible field-effect device. *Sci. Adv.* **2019**, *5*, eaav7282. [CrossRef]
119. Motoyama, E.M.; Yu, G.; Vishik, I.M.; Vajk, O.P.; Mang, P.K.; Greven, M. Spin correlations in the electron-doped high-transition-temperature superconductor $Nd_{2-x}Ce_xCuO_{4\pm\delta}$. *Nature* **2007**, *445*, 186–189. [CrossRef] [PubMed]
120. Yamada, K.; Kurahashi, K.; Uefuji, T.; Fujita, M.; Park, S.; Lee, S.H.; Endoh, Y. Commensurate Spin Dynamics in the Superconducting State of an Electron-Doped Cuprate Superconductor. *Phys. Rev. Lett.* **2003**, *90*, 137004. [CrossRef]
121. Zhao, Y.; Gaulin, B.D.; Castellan, J.P.; Ruff, J.P.C.; Dunsiger, S.R.; Gu, G.D.; Dabkowska, H.A. High-resolution x-ray scattering studies of structural phase transitions in underdoped $La_{2-x}Ba_xCuO_4$. *Phys. Rev. B* **2007**, *76*, 184121. [CrossRef]
122. Yu, G.; Li, Y.; Motoyama, E.M.; Zhao, X.; Barišić, N.; Cho, Y.; Bourges, P.; Hradil, K.; Mole, R.A.; Greven, M. Magnetic resonance in the model high-temperature superconductor $HgBa_2CuO_{4+\delta}$. *Phys. Rev. B* **2010**, *81*, 064518. [CrossRef]

symmetry

MDPI

Article

Electronic Dislocation Dynamics in Metastable Wigner Crystal States

Andrej Kranjec [1], Petr Karpov [2,3], Yevhenii Vaskivskyi [1], Jaka Vodeb [1], Yaroslav Gerasimenko [1] and Dragan Mihailovic [1,4,*]

1 Jozef Stefan Institute, Jamova 39, SI-1000 Ljubljana, Slovenia; andrej.kranjec@ijs.si (A.K.); yevhenii.vaskivskyi@ijs.si (Y.V.); jaka.vodeb@ijs.si (J.V.); yaroslav.gerasimenko@ijs.si (Y.G.)
2 Max Planck Institute for Physics of Complex Systems, Nöthnitzer Str. 38, 01187 Dresden, Germany; karpov.petr@gmail.com
3 Arnold Sommerfeld Center for Theoretical Physics, Ludwig-Maximilians-Universität München, Theresienstr. 37, 80333 Munich, Germany
4 CENN Nanocenter, Jamova 39, SI-1000 Ljubljana, Slovenia
* Correspondence: dragan.mihailovic@ijs.si

Abstract: Metastable states appear in many areas of physics as a result of symmetry-breaking phase transitions. An important challenge is to understand the microscopic mechanisms which lead to the formation of the energy barrier separating a metastable state from the ground state. In this paper, we describe an experimental example of the hidden metastable domain state in 1T-TaS_2, created by photoexcitation or carrier injection. The system is an example of a charge density wave superlattice in the Wigner crystal limit displaying discommensurations and domain formation when additional charge is injected either through contacts or by photoexcitation. The domain walls and their crossings in particular display interesting, topologically entangled structures, which have a crucial role in the metastability of the system. We model the properties of experimentally observed thermally activated dynamics of topologically protected defects—dislocations—whose annihilation dynamics can be observed experimentally by scanning tunnelling microscopy as emergent phenomena described by a doped Wigner crystal. The different dynamics of trivial and non-trivial topological defects are quite striking. Trivial defects appear to annihilate quite rapidly at low temperatures on the timescale of the experiments, while non-trivial defects annihilate rarely, if at all.

Keywords: domain walls; electronic crystal; Wigner crystal; polarons; TaS_2; transition metal dichalcogenide; topological defect; dislocations; scanning tunnelling microscopy; topological protection

1. Introduction

Metallic transition metal dichalcogenides (TMDs) commonly display commensurate charge-ordering at low temperatures. While some materials have received particularly detailed attention with a focus on specific features, the generic features, and the origin of broken translational symmetry, in particular, cannot be easily described in terms of traditional concepts, such as Fermi surface nesting, exciton condensation phenomena or peculiarities of the electron–phonon interaction occurring in specific systems. On the other hand, a lattice Wigner crystal description appears to offer a common conceptual starting point for discussion not only of the ground-state ordering, but also metastable states in the system. The justification for the lattice Wigner crystal comes from the ratio of the kinetic energy T in relation to the Coulomb energy V, expressed in terms of the Wigner–Seitz radius r_s, that defines the crossover from Fermi-liquid behaviour to a lattice Wigner crystal state with dominant Coulomb interactions. The ratio of Coulomb energy to kinetic energy defines $r_s = \frac{V}{T} = e^2m^*/\left(\hbar^2 n^{1/2}\right)$, where n is the electron density, e is the elementary charge, and m^* is the effective mass of electron [1]. r_s is strongly dependent on the electron density n. In the prototypical dichalcogenide charge-density-wave (CDW) system—1T-TaS_2, for example—the electron density n is small (one electron per 13 unit cells). Due to strong coupling to the lattice [2], polaronic effects renormalize the electronic bands, which become narrow, resulting in an increase in effective mass [3], which leads to an $r_s = 70 \sim 100$ and carrier localization in the lattice Wigner-crystal limit. There are two consequences: the first is the introduction of Mott physics [4,5], where the ordering dynamics are dominated by the Coulomb interaction, and the second is instability towards domain formation upon external perturbation or doping that takes place in the form of a symmetry-breaking transition.

The domains in such electronic crystals display very rich physics [6–9]. For example, the structure of domains created by laser excitation displays intertwined orders and intricate chiral ordering [9]. The structure of domain walls (DWs) themselves has been addressed on different levels [10–13]. In the context of Wigner crystal domains, Karpov and Brazovskii [10] introduced fractional charges associated with different mutual domain configurations. On a microscopic level, domain walls were analysed with scanning tunnelling microscopy in combination with density-functional-theory calculations [10–13], revealing the electronic structure within DWs. However, such modelling cannot be used to understand domain formation kinetics, the origins of metastability [14], or the dynamics of domain relaxation, governed by spatial topological constraints; nor can it be used to predict the generic formation of commensurate and domain structures across the entire series of metallic TMDs. The first step toward understanding such phenomena is the experimental characterization of the topology of the domain structure.

Here, we present an experimental analysis of the domain wall junctions, analysing the vertex structure in terms of 2D classical edge dislocations. Real-time measurements of domain relaxation using scanning tunnelling microscopy show the possibility of measuring dislocation annihilation dynamics.

2. Experimental Results

At low temperature (<140 K), the unperturbed ground state in 1T-TaS_2 is an electronic superlattice with a filling of 1/13 with respect to the atomic triangular lattice, which is commonly referred to as the commensurate charge-density-wave (CCDW or C-state). The superlattice vector is given by $\boldsymbol{A} = 3\boldsymbol{a} + \boldsymbol{b}$, where $\boldsymbol{a}$ and $\boldsymbol{b}$ are the fundamental crystal lattice vectors (Figure 1a), the stars signify the polaronic distortions around extra electron charge on the Ta atom at the centre of each star-of-David hexagram. Domains are ubiquitously observed when a charge is injected by a large STM tip current pulse [6], an electrical pulse through contacts of a device [8], or by photoexcitation (photodoping), where the effect was originally discovered [14]. The resulting 'hidden' (H) domain state is metastable and has a lifetime that is sufficiently long that its domain structure can be investigated in detail at low temperatures [15].

In the present work, we excite in situ cleaved single crystals of 1T-TaS_2 at 4.2 K with a voltage pulse from the tungsten tip of an STM. The area around the location of the voltage pulse is first probed to determine where the hidden state borders the surrounding CCDW. When an appropriate area is found, we let the STM scanning run until at least several frames are acquired and a substantial relaxation is observed. Typical domain-forming pulse parameters are V = 4 V, pulse duration 1 ms and tunnel junction resistance 1 GΩ.

To study the topological properties of the domain network, we need to accurately determine domains' phase shifts, fitting them on the underlying atomic lattice. To this end, we employ a simulated annealing Monte Carlo (MC) algorithm: first, we determine the most probable values of phase shifts for each neighbouring pair of domains, which we then globally optimize for the whole domain network. Such a local procedure supplemented by global "sewing" helps us to determine the domains' phases, even for big STM images which exhibit non-uniform values of the underlying lattice constant or lattice rotation angle.

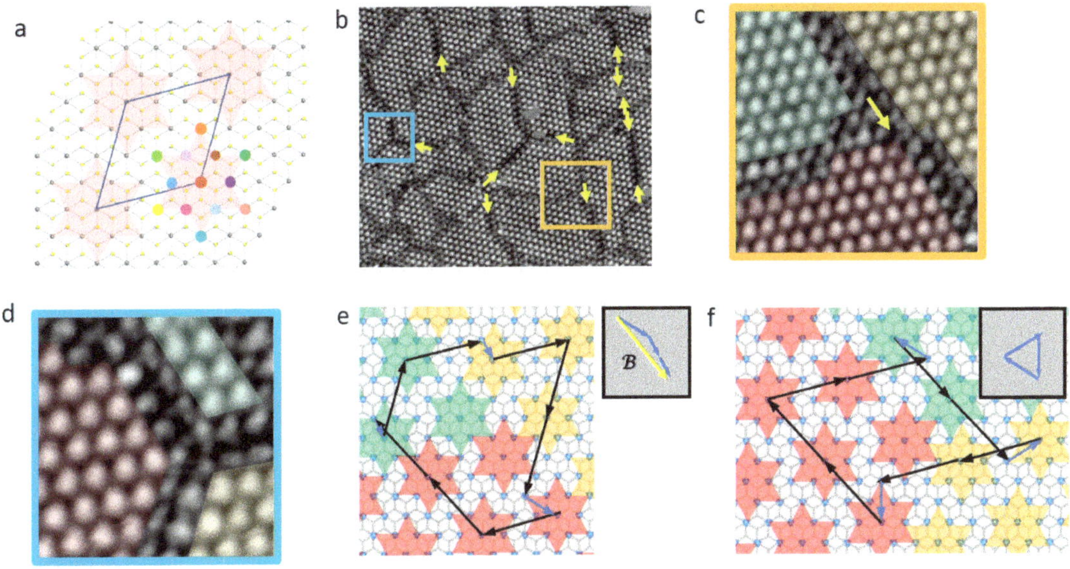

Figure 1. (**a**) Schematic representation of the commensurate superlattice showing twelve possible interdomain phase shifts with the corresponding colour coding. The parallelogram shows the CDW superlattice unit cell. (**b**) STM image (V = −800 mV) of a part of the domain state with a relatively high density of dislocations. Yellow arrows indicate non-zero Burgers vectors at the vertices. (**c**) Zoom-in of the orange square on (**b**), rotated to align the underlying atomic lattice with the horizontal axis. Colour shading indicates different domains. A yellow arrow represents the Burgers vector. (**d**) Same as in (**c**) for a trivial vertex marked with the blue square in (**b**). (**e**) A schematic representation of the 3-vertex in (**c**). Blue circles are tantalum atoms; star of David overlays represent polarons using the same colour scheme as (**c**). Black arrows are CDW superlattice vectors indicating the Burgers circuit around the vertex. Blue vectors are relative shifts of domains upon crossing a domain wall. The inset shows that the relative shifts add up to a CDW superlattice vector, the Burgers vector of the vertex. (**f**) Same as in (**e**) for the trivial vertex. The inset of image 9 (**t**) shows that relative shift vectors add up to zero. Note that all domain wall types are the most common ones observed as reported previously in Ref. [13].

Using the above procedure, we analyse a large area domain structure formed by STM tip current injection shown in Figure 1b. The domain walls cross, typically forming three or four-pronged vertices. The vertices may be trivial, or non-trivial, according to whether a Burger's vector construction around the vertex is zero ($\mathcal{B} = 0$), or non-zero ($\mathcal{B} \neq 0$), to which we refer as trivial (t) and non-trivial (nt) vertices, respectively. In Figure 1b, vertices are labelled according to whether the Burger's vector construction is t (no arrow), or nt (the yellow arrow shows $\mathcal{B}$). Figure 1c,d show examples of nt and a t vertices, respectively. The corresponding Burgers circuit construction in each case is shown in Figure 1e,f. The construction is always made, along with the shortest possible interdomain shifts upon crossing a domain wall, which is either of the nearest-neighbour types, with length $\boldsymbol{a}$, or the next nearest-neighbour type with length $\sqrt{3}\boldsymbol{a}$ (Figure 1e,f). We emphasize that the Burgers vector construction does not depend on the detailed structure of the vertex or the presence of extrinsic defects in its vicinity.

A relaxation sequence of the H state induced by a voltage pulse from the STM tip is shown in Figure 2. The domain area is entirely surrounded by the C-state (red overlay). The panels a–b6 show a gradual diminishing of the H-state and expansion of the C-state in discrete steps. Each frame was analysed by the MC-simulated annealing algorithm to determine the presence of dislocations at vertices. The vertices are labelled according to how many domains are adjacent to the vertex, which is the first number, and whether they are trivial (*t*) or non-trivial (*nt*). The arrows next to non-trivial vertices indicate their Burgers vector, yellow for 4*nt* and white for 3*nt*. Panels b1–b6, with overlaid Burgers vectors, display the dislocation dynamics. We observe a constant number of dislocations in the domain reconfiguration process until the last frame, where two dislocations with opposite Burgers vectors annihilate. The latter two are marked by blue circles on image b5 prior to annihilation. Movement of some dislocations is also observed (which we attribute to a combination of climb and glide—see Figure 3).

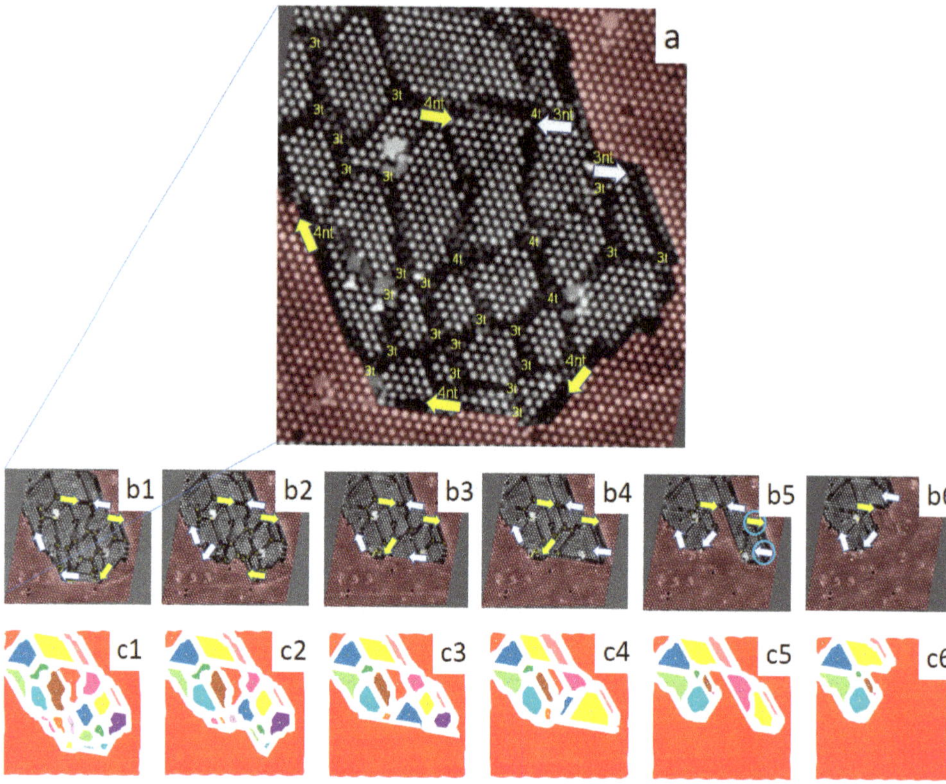

Figure 2. Relaxation sequence of the hidden state induced by a voltage pulse from the STM tip. (**a,b**) Analysed STM sequence of the relaxation of a patch of hidden state surrounded by the CCDW. Vertices are marked by yellow markings, with the number representing the number of domains that meet at a vertex and the letters "*t*" and "*nt*" stand for trivial and non-trivial, respectively (3*nt* = non-trivial 3-vertex). The yellow arrow next to non-trivial vertices indicate their Burgers vector. Red shading indicates the surrounding CCDW and the domain, which is not shifted with respect to the CCDW. Blue circles in frames 4- and 5-mark pairs of dislocations that annihilate in the next frame. (**c**) Colour maps of domains for each frame in (**a,b**) as a result of the Monte Carlo algorithm. Colours correspond to relative domain shifts according to Figure 1d, with the origin (red) chosen as the CCDW. The colour of lines connecting centres of domains represents the quality of recognition of the MC algorithm. Green edges correspond to the best reliability, orange to lesser, red even less and black poor recognition. The vast majority of lines in all frames are green indicating reliable interdomain shift recognition. White lines in the CCDW are used to cut it into smaller pieces for better algorithm recognition.

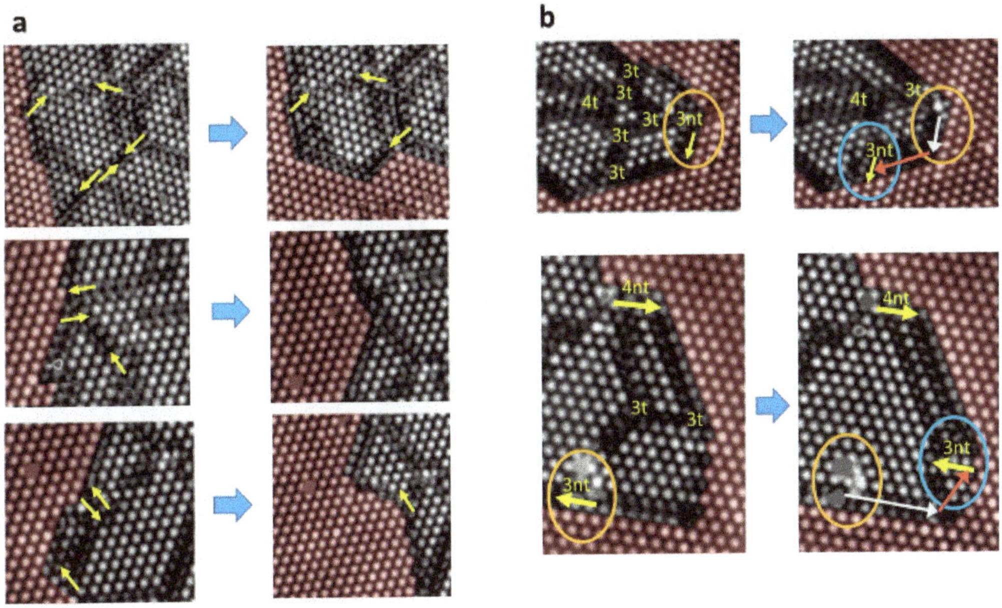

Figure 3. (**a**) Relaxation events involving annihilation of non-trivial defects. The Burger's vectors are indicated by the yellow arrows; (**b**) two examples of glide and climb motion of 3*nt* dislocations. The red overlay represents the bordering CCDW.

The panels c1–c6 show the same time evolution with colour maps for domains generated by the MC algorithm. The colouring of domains is according to their shift with respect to the original reference domain (which is chosen to be the C-state) following the colouring scheme of Figure 1a. We note that the colouring of domains remains consistent throughout the relaxation in the sense that domains do not change the CDW phase until they disappear. This is to be expected, since it is energetically costly to change the phase shift of a whole domain. They do, however, change their shape and size during relaxation.

Annihilation, glide and climb. Figure 3a shows three examples of dislocation pair annihilation in subsequent frames. In the middle panel, three dislocations disappear; in the last panel, two dislocations annihilate, and one is displaced. Yellow arrows are Burgers vectors, red shading indicates the surrounding CCDW. Figure 3b shows two examples of dislocation (3*nt* vertex) displacements, which we can attribute to a combination of glide and climb motion, as indicated by the different-coloured arrows (glide: white, climb: red). Glide motion involves motion along the direction of $\boldsymbol{\mathcal{B}}$, while climb involves motion at an angle of either 60° or 120° to $\boldsymbol{\mathcal{B}}$.

Annihilation dynamics. By examining STM frames in sequence, we can obtain an idea of the different rates of dislocation annihilation. While the amount of data is limited by the small number of frames and the slow relaxation rate, we can clearly distinguish different relaxation rates for different types of vertices. Figure 4 shows a count of the number of *t* and *nt* vertices for three- and four-pronged vertices on the full STM sequence of images shown in Figure 2. The three-pronged *t* vertices appear to show relaxation on the timescale of tens of minutes. In contrast, the four-pronged vertices show much slower dynamics. This is possibly due to the fact that most 3*t* vertices are located closer to the border with the CCDW, or simply due to the fact that they have more phase constraints and are therefore more energetically costly to relax. The *nt* defects show very little dynamics on this timescale. The number of dislocations is constant until the last frame (when two dislocations annihilate) and so they form a sort of stable backbone of the domain wall network. It is clear from the examination of the STM images that t and nt vertices are inextricably entangled with each other in the sense that vertex $(3nt)$ and anti-vertex $(\overline{3nt})$ pairs are not adjacent to each other, but there may be other vertices geometrically constraining their annihilation. It is quite likely that some defects have their anti-defects outside the image area, and the present experiments are limited in this way. Nevertheless, the general trend is clear. We also note that the *fraction* of nt vertices increases during relaxation, making the hidden state more topologically protected. This could have implications for data retention in memory-device applications.

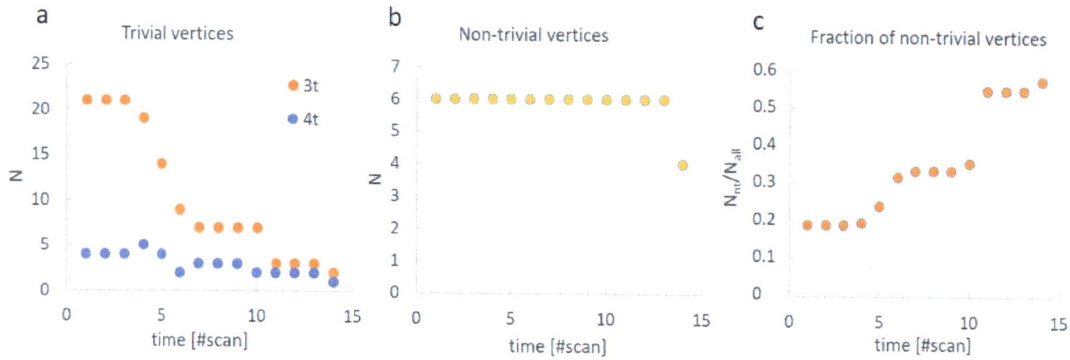

Figure 4. Temporal dynamics of (**a**) trivial (t) and (**b**) non-trivial (nt) vertex relaxation, respectively in consecutive STM scans. Each scan takes around 20 min. (Note the different scales.) (**c**) The number of nt vertices as a fraction of the total number of vertices.

3. Domain Textures as an Emergent Phenomenon

To understand the formation of domains, one may approach it from different directions [9–14,16,17]. A generally applicable theory with useful predictive value considers the domain structure as an emergent property, as a response of the polaronic Wigner crystal to doping [10,14,16]. The model Hamiltonian is given by [16]:

$$H = \Sigma_{I,j} V_{i,j} n_i n_j \quad (1)$$

where $n_i = 0$ or 1 is the occupation number of a polaron on the site i, and

$$V(i, j) = V_0\, exp(-r_{ij}/r_s)/r_{ij} \quad (2)$$

is the Yukawa screening potential, where $V_0 = e^2/\epsilon_0 a$ in CGS units and $r_{ij} = |r_i - r_j|$, where r_i is the dimensionless position of the i-th polaron, and r_s is the dimensionless screening radius. ϵ_0 is the dielectric constant of the material. The use of the model for quasi-2D transition metal dichalcogenides has been justified previously, and has predicted the existence of electronic superlattices at magic-filling fractions and domains or amorphous electronic structures in between. Details of the calculation can be found in Ref. [16].

One possible way to represent the topological defects in the data is by using Wigner–Seitz cell (or sometimes called Voronoi) constructions around each polaron. Figure 5b shows such a construction for the large area experimental STM image with a large density of dislocations in Figure 5a. As before, the domains are first put to the underlying lattice with the MC simulated annealing process. Hexagons are coloured blue, while pentagons and heptagons are coloured orange and green, respectively. We see that nt vertices indicated by the arrows (which have non-zero $\mathcal{B}$) are associated with heptagon–pentagon pairs. These are topologically equivalent to dislocations on a hexagonal lattice. The Wigner–Seitz cell representation is thus a useful abstraction for the rather complicated domain structure observed in STM experiments.

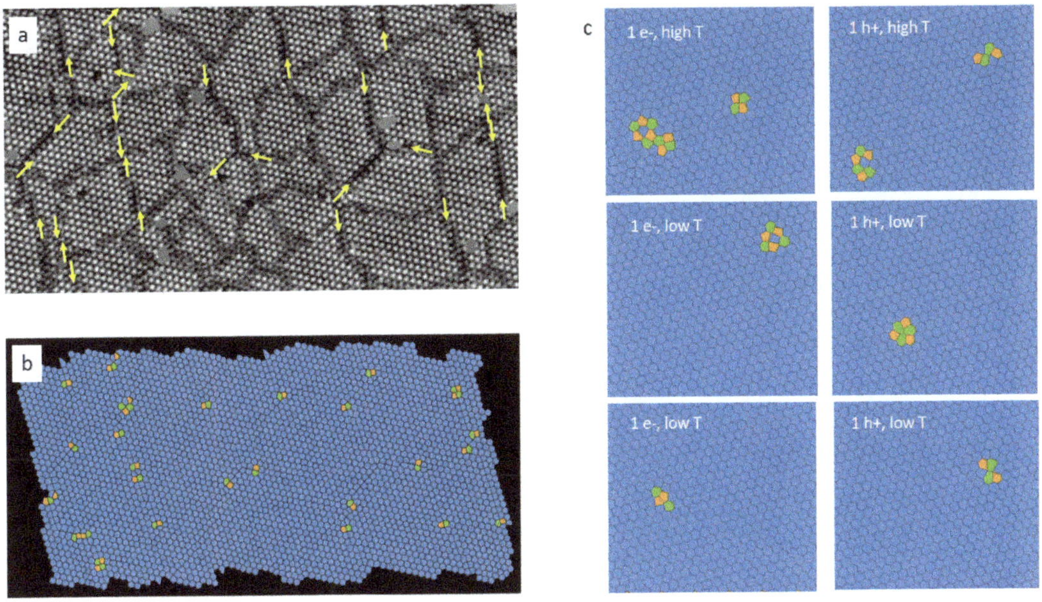

Figure 5. Comparison of an experimental image with the MC lattice gas model. (**a**) An STM image with a relatively high density of dislocations. Yellow arrows are Burgers vectors. (**b**) Wigner–Seitz cell diagram of the MC lattice fit to the image in (**a**). (**c**) Wigner–Seitz cells of the MC lattice gas model simulation after adding an electron (e-), or a hole (h+) to the lattice at different effective temperatures. Colour code: hexagons (blue), pentagons (orange), and heptagons (green).

The results of the calculation for lightly doped (adding 1 electron or 1 hole) Wigner crystal with a magic filling fraction of 1/13 are shown in Figure 5c in the form of a Wigner–Seitz cell plot. We see that doping creates topological defects in the form of pentagon–heptagon pairs, irrespective of the temperature. At higher temperatures, additional defects are visible due to $e - h$ pair excitations appearing as fluctuations in the number density. Comparing the predictions of this classical polaronic lattice gas model (Figure 5c) with the experiment in Figure 5b, we see that the model predicts the formation of the same kind of defects associated with doping. In both cases, we can see paired dislocations. The difference is in the presence of single heptagon–pentagon pairs in the experimental image, while in the model they are all paired or clustered. The comparison strongly suggests that the vertices are associated with an extra trapped charge.

Our simulations also show that the number of defects increases as we increase doping in the system and when doping is zero, no defects are remaining. Therefore, from the point of view of our polaronic Wigner crystal approach, the domain formation mechanism is the following. Photoexcitation or charge injection introduces additional charges in the system. Karpov and Brazovskii [10] have already shown that a domain wall is energetically more favourable than a hole or interstitial charge; therefore, a network of domain walls forms. The number of topologically protected defects is governed by the doping level and this number does not change if the doping level remains the same. Trivial defects, however, can relax and disappear. Experimentally, it was noted before [16] that the doping level differs in the ground state and metastable state, which suggests that extra charges responsible for topologically protected defects also eventually leak back into the Fermi sea below the Fermi level.

4. Discussion

'Standard' theory of domain walls in CDW systems considers the energy of DW formation and energy of DW crossings (DWCs), which are of opposite signs, such that the two form a metastable minimum in the free energy of the system [14,18–21]. The minimum corresponds to a particular, topologically protected domain state, in the sense that the minimum is determined by specific details of the material and is immune to external single-particle perturbations. However, this does not include the additional complication of real-space pinning defects, nor is it immune to strain, for example [15]. In the same sense, the domain size is controlled by carrier injection [14,22,23]. The self-organized intrinsic formation of topological defects at DW crossings observed above is not included in the theory, yet it is likely to have a significant effect on the system stability, and hence the free-energy potential landscape. The dynamics of dislocations in particular are governed by annihilation dynamics, the bottleneck being matched defect pair annihilation. In 2D, conventional Kosterlitz–Thouless (K-T) theory, the dynamics are discussed in terms of edge dislocations, which are created and annihilated in pairs. At high density, the dislocations destroy long-range order, resulting in a hexatic phase. When the dislocations dissociate into five- and seven-fold disclinations, an additional isotropic phase may appear. The two-step melting of such a phase is described in terms of the Kosterlitz–Thouless–Halperin–Nelson–Young (KTHNY) theory with two consecutive K-T transitions. These theories do not consider the dynamics of the dislocations, which can be quite intricate.

The dynamics of classical lattice dislocations under stress is governed by a Peach–Koehler (P-K) force between dislocations described in terms of their Burger's vector and the stress tensor, with additional terms arising from pinning by extrinsic defects, temperature-driven lattice fluctuations, local (Peierls-Nabarro) strains, etc. [24,25]. Simulations of the dynamics in classical 2D systems typically reveal aggregation of dislocations in the form of clusters, and at the edges.

In the case of an electronic crystal on a hexagonal lattice, the structure of non-trivial defects is formally equivalent to lattice dislocations, but the microscopic structural details, on which the dynamics will depend, are quite specific. The experimental data above show the structure of trivial and non-trivial vertex dynamics and are presented formally in terms of dislocation dynamics. Aggregation of dislocations at the edges or in the centre of the domain structures is not observed on the timescale of the STM experiments, suggesting that annihilation is faster than the P-K dynamics at 4 K. More importantly, the dynamics is governed by the entanglement of defect pairs, which prevents direct annihilation of defect–anti-defect pairs through a simple glide or climb motion, but requires 'circumnavigation' and annihilation in a particular sequence. The fact that the domains are created spontaneously, from initial fluctuation–nucleation after a quench suggests some degree of randomness in the initial distribution of defects. However, domain-wall repulsion and the energy gained by crossings at vertices act to form a more uniform, long-ranged order in the domain structure. The latter process is hindered by topological constraints at the vertices.

We note that the present analysis is limited to the temporal and spatial window of the STM experiments. Dynamics on other timescales cannot be excluded, but currently, there are no other methods to observe the defect dynamics. In spite of experimental limitations, the dynamics of the observed difference in *nt* and *t* dynamics are quite significant, indicating that the forces between *nt* and *t* are quite different. An additional complication arises from the fact that *nt* and *t* vertices are intertwined, which brings a complex topology to the problem.

On a phenomenological level, the relaxation was previously described in terms of the Ostwald ripening process, with double exponential relaxation, but the approach is not very insightful in explaining the magnitude of the energy barrier E_B for relaxation to the ground state [15]. However, the strong observed sensitivity of E_B to external strain suggests that a general P-K approach is reasonable.

A suitable framework for describing the quantum nature of such a system may be provided by fractions, excitations within a tensorial field theory that are dual to classical elasticity theory used in the present analysis [26], but such treatment is beyond the scope of this paper.

Given that the detailed construction of DWCs is likely to play an important role in the overall relaxation process of the electronic Wigner crystal, one may comment on the prospect of dealing with the dynamics of such a system on the microscopic level (e.g., density functional theory) used recently to describe the static structure of single DWs. Such modelling may be useful in determining the relative energy stability of different DWs and eventually DWCs. The transitions between states on the resulting energy landscape may form the basis of more material-specific relaxation dynamics.

5. Conclusions

The difference in t and nt dynamics is striking and proves that the topological protection of nt vertices plays a crucial role in the stability of the domain state. We conclude by commenting on the fact that at low temperatures, the dynamics are limited by macroscopic quantum tunnelling between different, but energetically near-degenerate domain configurations that were shown, in recent experiments, to be finite [27].

Author Contributions: Conceptualization, D.M., A.K., Y.G. and Y.V.; methodology, P.K., A.K. and D.M.; software, P.K., A.K. and J.V.; formal analysis, A.K., P.K. and J.V.; writing—original draft preparation, D.M. and A.K.; writing—review and editing, D.M., A.K., P.K., Y.V. and J.V.; funding acquisition, D.M. All authors have read and agreed to the published version of the manuscript.

Funding: This research was funded by the Slovenian Research Agency, grant P1-0040 and A.K. to PR-06158. We thank the CENN Nanocenter for the use of its facilities.

Data Availability Statement: Not applicable.

Acknowledgments: We thank the CENN Nanocenter for the use of its facilities, Petra Šutar for the samples used in this work, and Maksim Litskevich for the help with measurements.

Conflicts of Interest: The authors declare no conflict of interest.

References

1. Schrieffer, J.R. *Theory of Superconductivity*; Westview Books: Boulder, CO, USA, 1964; ISBN 0738201200.
2. Machida, Y.; Hanashima, T.; Ohkubo, K.; Yamawaki, K.; Tanaka, M.; Sasaki, S. Observation of Soft Phonon Modes in 1T-TaS 2by Means of X-ray Thermal Diffuse Scattering. *J. Phys. Soc. Jpn.* **2004**, *73*, 3064–3069. [CrossRef]
3. Rossnagel, K. On the Origin of Charge-Density Waves in Select Layered Transition-Metal Dichalcogenides. *J. Phys. Condens. Matter* **2011**, *23*, 213001. [CrossRef] [PubMed]
4. Fazekas, P.; Tosatti, E. Charge Carrier Localization in Pure and Doped 1T-Tas2. *Phys. B+C* **1980**, *99*, 183–187. [CrossRef]
5. Tosatti, E.; Fazekas, P. On the Nature of the Low-Temperature Phase of 1T-TaS2. *J. Phys. Colloq.* **1976**, *37*, C4-165–C4-168. [CrossRef]
6. Ma, L.; Ye, C.; Yu, Y.; Lu, X.F.; Niu, X.; Kim, S.; Feng, D.; Tománek, D.; Son, Y.W.; Chen, X.H.; et al. A Metallic Mosaic Phase and the Origin of Mott-Insulating State in 1T-TaS2. *Nat. Commun.* **2016**, *7*, 10956. [CrossRef]
7. Cho, D.; Cheon, S.; Kim, K.-S.; Lee, S.-H.; Cho, Y.-H.; Cheong, S.-W.; Yeom, H.W. Nanoscale Manipulation of the Mott Insulating State Coupled to Charge Order in 1T-TaS2. *Nat. Commun.* **2016**, *7*, 10453. [CrossRef]
8. Vaskivskyi, I.; Mihailovic, I.A.; Brazovskii, S.; Gospodaric, J.; Mertelj, T.; Svetin, D.; Sutar, P.; Mihailovic, D. Fast Electronic Resistance Switching Involving Hidden Charge Density Wave States. *Nat. Commun.* **2016**, *7*, 11442. [CrossRef]
9. Gerasimenko, Y.A.; Karpov, P.; Vaskivskyi, I.; Brazovskii, S.; Mihailovic, D. Intertwined Chiral Charge Orders and Topological Stabilization of the Light-Induced State of a Prototypical Transition Metal Dichalcogenide. *npj Quantum Mater.* **2019**, *4*, 32. [CrossRef]
10. Karpov, P.; Brazovskii, S. Modeling of Networks and Globules of Charged Domain Walls Observed in Pump and Pulse Induced States. *Sci. Rep.* **2018**, *8*, 4043. [CrossRef]
11. Cho, D.; Gye, G.; Lee, J.; Lee, S.-H.; Wang, L.; Cheong, S.-W.; Yeom, H.W. Correlated Electronic States at Domain Walls of a Mott-Charge-Density-Wave Insulator 1 T-TaS 2. *Nat. Commun.* **2017**, *8*, 392. [CrossRef]
12. Park, J.W.; Cho, G.Y.; Lee, J.; Yeom, H.W. Emergent Honeycomb Network of Topological Excitations in Correlated Charge Density Wave. *Nat. Commun.* **2019**, *10*, 4038. [CrossRef] [PubMed]

13. Park, J.W.; Lee, J.; Yeom, H.W. Zoology of Domain Walls in Quasi-2D Correlated Charge Density Wave of 1T-TaS2. *NPJ Quantum Mater.* **2021**, *6*, 32. [CrossRef]
14. Stojchevska, L.; Vaskivskyi, I.; Mertelj, T.; Kusar, P.; Svetin, D.; Brazovskii, S.; Mihailovic, D. Ultrafast Switching to a Stable Hidden Quantum State in an Electronic Crystal. *Science* **2014**, *344*, 177–180. [CrossRef] [PubMed]
15. Vaskivskyi, I.; Gospodaric, J.; Brazovskii, S.; Svetin, D.; Sutar, P.; Goreshnik, E.; Mihailovic, I.A.; Mertelj, T.; Mihailović, D. Controlling the Metal-to-Insulator Relaxation of the Metastable Hidden Quantum State in 1T-TaS2. *Sci. Adv.* **2015**, *1*, e1500168. [CrossRef]
16. Vodeb, J.; Kabanov, V.V.; Gerasimenko, Y.A.; Venturini, R.; Ravnik, J.; van Midden, M.A.; Zupanic, E.; Sutar, P.; Mihailovic, D. Configurational Electronic States in Layered Transition Metal Dichalcogenides. *New J. Phys.* **2019**, *21*, 083001. [CrossRef]
17. Ravnik, J.; Diego, M.; Gerasimenko, Y.; Vaskivskyi, Y.; Vaskivskyi, I.; Mertelj, T.; Vodeb, J.; Mihailovic, D. A Time-Domain Phase Diagram of Metastable States in a Charge Ordered Quantum Material. *Nat. Commun.* **2021**, *12*, 2323. [CrossRef]
18. Villain, J. Commensurate-Incommensurate Transition of Krypton Monolayers on Graphite: A Low Temperature Theory. *Surf. Sci.* **1980**, *97*, 219–242. [CrossRef]
19. Bak, P.; Mukamel, D.; Villain, J.; Wentowska, K. Commensurate-Incommensurate Transitions in Rare-Gas Monolayers Adsorbed on Graphite and in Layered Charge-Density-Wave Systems. *Phys. Rev. B* **1979**, *19*, 1610. [CrossRef]
20. Bak, P. Commensurate Phases, Incommensurate Phases and the Devil's Staircase. *Rep. Prog. Phys.* **1982**, *45*, 587–629. [CrossRef]
21. Brazovskii, S. Modeling of Evolution of a Complex Electronic System to an Ordered Hidden State: Application to Optical Quench in 1T-TaS2. *J. Supercond Nov. Magn.* **2015**, *28*, 1349–1353. [CrossRef]
22. Brazovskii, S. Coherent Topological Defect Dynamics and Collective Modes in Superconductors and Electronic Crystals. Available online: http://iopscience.iop.org/0953-8984/25/40/404206/pdf/0953-8984_25_40_404206.pdf (accessed on 12 March 2022).
23. Brazovskii, S.; Kirova, N.; Requardt, H.; Nad, F.Y.; Monceau, P.; Currat, R.; Lorenzo, J.E.; Grübel, G.; Vettier, C. Plastic Sliding of Charge Density Waves: X-ray Space Resolved-Studies versus Theory of Current Conversion. *Phys. Rev. B* **2000**, *61*, 10640. [CrossRef]
24. Roters, F.; Eisenlohr, P.; Hantcherli, L.; Tjahjanto, D.D.; Bieler, T.R.; Raabe, D. Overview of Constitutive Laws, Kinematics, Homogenization and Multiscale Methods in Crystal Plasticity Finite-Element Modeling: Theory, Experiments, Applications. *Acta Mater.* **2010**, *58*, 1152–1211. [CrossRef]
25. Lubarda, V.A. Dislocation Burgers Vector and the Peach–Koehler Force: A Review. *J. Mater. Res. Technol.* **2019**, *8*, 1550–1565. [CrossRef]
26. Pretko, M.; Radzihovsky, L. Fracton-Elasticity Duality. *Phys. Rev. Lett.* **2018**, *120*, 195301. [CrossRef]
27. Vodeb, J.; Diego, M.; Vaskivskyi, Y.; Gerasimenko, Y.; Kabanov, V.; Mihailovic, D. Observation of Quantum Domain Melting and Its Simulation with a Quantum Computer. *arXiv* **2021**, arXiv:2103.07343.

symmetry MDPI

Perspective

Scanning Probe Microscopy Investigation of Topological Defects

Jan Seidel [1,2]

[1] School of Materials Science and Engineering, The University of New South Wales, Sydney, NSW 2052, Australia; jan.seidel@unsw.edu.au

[2] ARC Centre of Excellence in Future Low-Energy Electronics Technologies, The University of New South Wales, Sydney, NSW 2052, Australia

Abstract: Symmetry lowering phase transitions in ferroelectrics, magnets, and materials with various other forms of inherent order lead to the formation of topological defects. Their non-trivial real-space topology is characterized by a topological charge, which represents the topological invariant. The study of topological defects in such materials has seen increased interest over the last decade. Among the methods used for their study, scanning probe microscopy (SPM) with its many variants has provided valuable new insight into these structures at the nanoscale. In this perspective, various approaches are discussed, and different techniques are compared with regard to their ability to investigate topological defect properties.

1. Introduction

Topological defect formation at the symmetry lowering phase transitions in solids with various forms of intrinsic order is a phenomenon widely studied in materials science and solid-state physics [1]. A prototypical example is a magnetic material cooled through its Curie temperature (see Figure 1), leading to the formation of magnetic domains and associated domain walls, which are a type of topological defect. From a general perspective, such nanoscale structures can have different intrinsic properties from the bulk material itself [2,3], making them interesting nanoscale objects with altered and additional functionality. These different properties are brought about by changes in local crystal structure, sometimes involving large structural gradients. In addition, local symmetry changes allow for the existence of properties forbidden in the higher symmetry bulk phase, again leading to changes from macroscopic materials properties.

These considerations have led to proposals of utilizing these altered properties at topological defects for enhanced material functionality, including for example domain wall nanoelectronics [4] and spintronic devices [5], which have been discussed in various contexts.

The structural width and size of such topological defects typical is found to be in the nanoscale range, down to atomic length scales. This small size requires high resolution characterization methods, involving electron microscopy, nanoscale spectroscopy methods, and in various forms, scanning probe microscopy, the latter being the focus of this article.

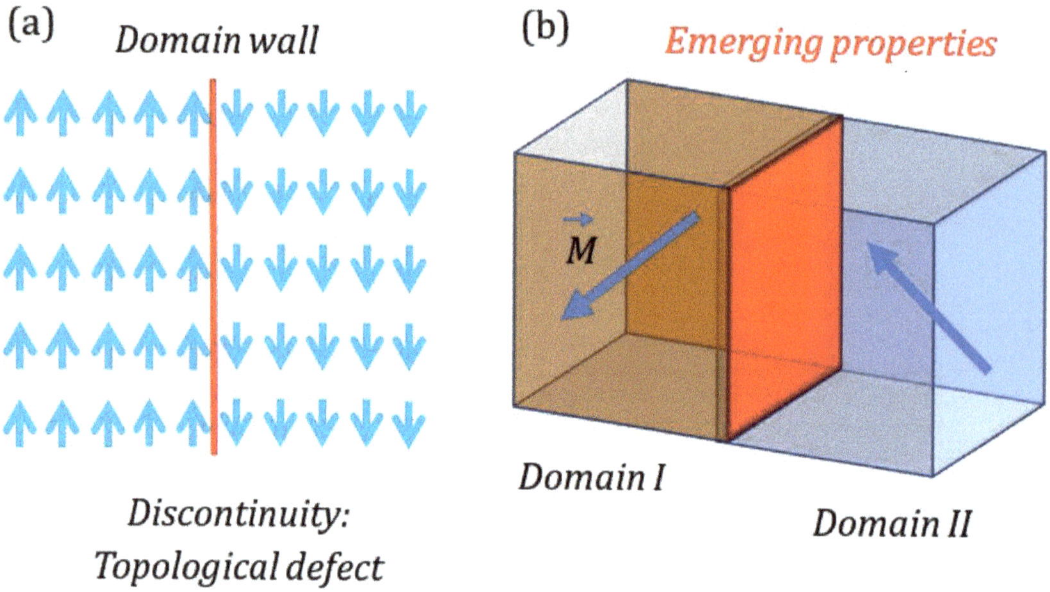

Figure 1. Schematic of a domain wall as an example of a topological defect. (**a**) The discontinuity formed in the built-in system order (for example magnetic moments, spins in magnetic materials) separates adjacent domains and can give rise to emerging properties; (**b**) emerging material properties at domain walls.

2. Types of Topological Defects

The types of topological defects found in solids [6,7] range from domain walls [4], skyrmions [8,9] and associated structures, vortices [10], to dislocations. These structures can be formed in various types of 'order background', involving magnetism (magnetic spins), ferroelectricty (electric dipoles), ferroelasticity (spontaneous strain), ferrotoroidicity, and crystal structure defects (dislocations). Their non-trivial real-space topology is characterized by a topological charge, which represents the topological invariant [11].

Domain walls in ferroelectric and multiferroic materials are a prominent example and have been investigated in more detail in the last decade. They have a very small size, on the order of a unit cell, as for example shown in Figure 2a for the case of a 109° domain wall in $BiFeO_3$. The colors in the image represent tilt angles of the perovskite unit cell overlaid on an atomic structure high-angle annular dark-field scanning transmission electron microscopy (HAADF-STEM) image. At the wall the unit cell tilt reaches ~3°, which is considerable if one takes into account its impact on bond lengths and orbital overlap in the ionic solid. These in turn lead to changes in electronic structure, and, hence the intrinsic properties are altered directly at the domain wall on the same atomic length scale.

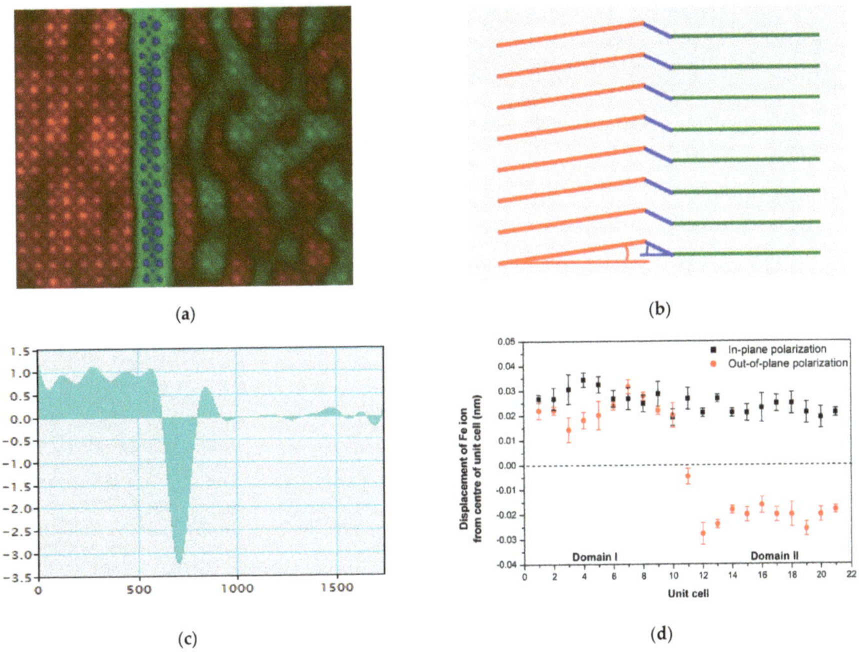

Figure 2. (**a**) False color HAADF-STEM image of a 109° domain wall in $BiFeO_3$; (**b**) unit cell tilt corresponding to colors in (**a**); (**c**) local unit cell tilt angle across the domain wall in (**a**); (**d**) Fe ion displacement depicting reorientation of ferroelectric polarization across the domain wall.

3. Scanning Probe Methods

Several different scanning probe methods have been used to study topological defects in solid materials, including their structural, electrical, magnetic, and other functional properties. Some methods are more suitable than others and the measured properties vary, as outlined in Table 1.

Table 1. Comparison of scanning probe methods used for the study of topological defects in various solid materials. Literature references point to relevant reviews of the technique, and examples of topological defect research, if applicable. Technique acronyms are explained in the text below.

Method	Lateral Resolution	Measured Quantity	Materials	Examples
AFM	atomic	surface morphology; mechanical properties	many, few restrictions	[12,13]
PFM	~5 nm	piezoresponse	ferro-/piezoelectrics	[14,15]
STM	atomic	tunnelling current	semiconductors, metals	[16,17]
KPFM	atomic	surface potential difference	semiconductors, metals	[18]
c-AFM	~5 nm	electrical conductivity	semiconductors, metals	[19,20]
Nano-IR	~10 nm	infrared optical properties	many with IR bands between 600–4500 cm^{-1}	[21]
NSOM	~10 nm	optical properties	many, few restrictions	[22,23]
MFM	~10 nm	magnetic field gradient	ferro-/ferrimagnets; superconductors	[24,25]
scanning SQUID	~100 nm	magnetic flux	antiferro-, ferro-/ferrimagnets; superconductors	[26]
scanning NV	~20 nm	magnetic field strength	antiferro-, ferro-/ferrimagnets; superconductors	[27,28]
sMIM	~50 nm	impedance	many, few restrictions	[29,30]
CGM	~5 nm	charge	ferro-/piezoelectrics	[31,32]

Application to Imaging and Manipulating Topological Defects

One feature that is common to all SPM methods is the acquisition of topographical information of measured sample surfaces, depicting their morphology. Even this basic atomic force microscopy (AFM) feature can be used for example to study twinning angles in ferroelastic twin domain walls [33] and to locate their position with sufficient accuracy for other local probing investigations, such as mechanical probing at the nanoscale [34].

Other techniques such as magnetic force microscopy (MFM) can be used to directly image magnetic domain structure, in magnetic materials. This method has also been used to directly visualize skyrmions [23] (Figure 3). Magnetic field dependent measurements show the 'melting' of the hexagonal skyrmion crystal lattice through the formation of magnetic monopoles. Other techniques to study magnetic properties include scanning SQUID microscopy [25], and scanning diamond color centre (scanning NV) microscopy [26,27]. The latter also allow for the study of antiferromagnets.

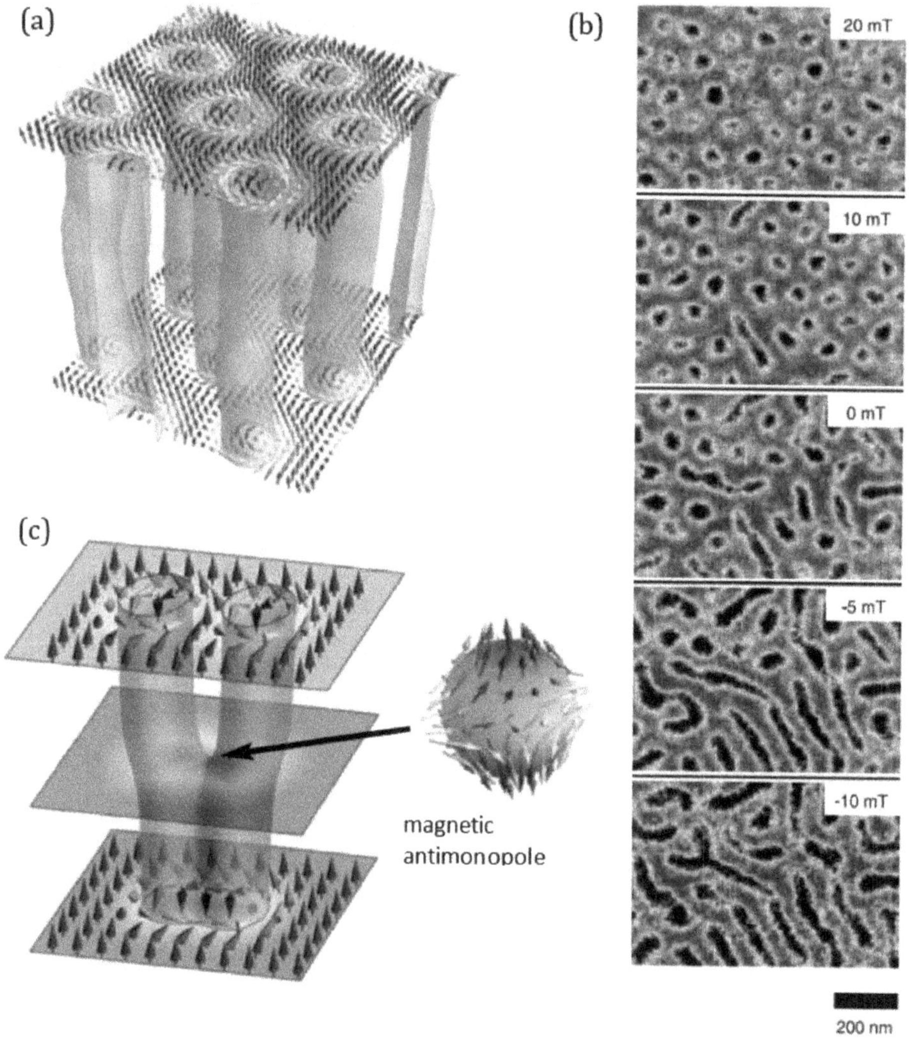

Figure 3. Magnetic force microscopy imaging of skyrmions in FeCoSi. (**a**) schematic skyrmion structure, (**b**) MFM image series in varying magnetic field showing the melting of the skyrmion crystal lattice, (**c**) 'unzipping' of skyrmion pairs involving a magnetic antimonopole.

Polar order in ferroelectrics can be visualized by piezoresponse force microscopy (PFM) and charge gradient microscopy (CGM), among others. This technique is often used in conjunction with other SPM modes for the investigation of topological defects, for example conductive AFM (c-AFM) to measure conductivity at domain walls in various ferroelectric and multiferroic oxides. The resolution in this case is given by the contact area of the SPM probe with the material surface (as is the case with several other techniques listed), and can be used to study domain wall devices as well as intrinsic material properties such as carrier properties, sometimes in conjunction with theoretical models. An example is shown in Figure 4, which shows a study of orthorhombic-rhombohedral phase boundaries (hybrid domain walls) in strained $BiFeO_3$ [35].

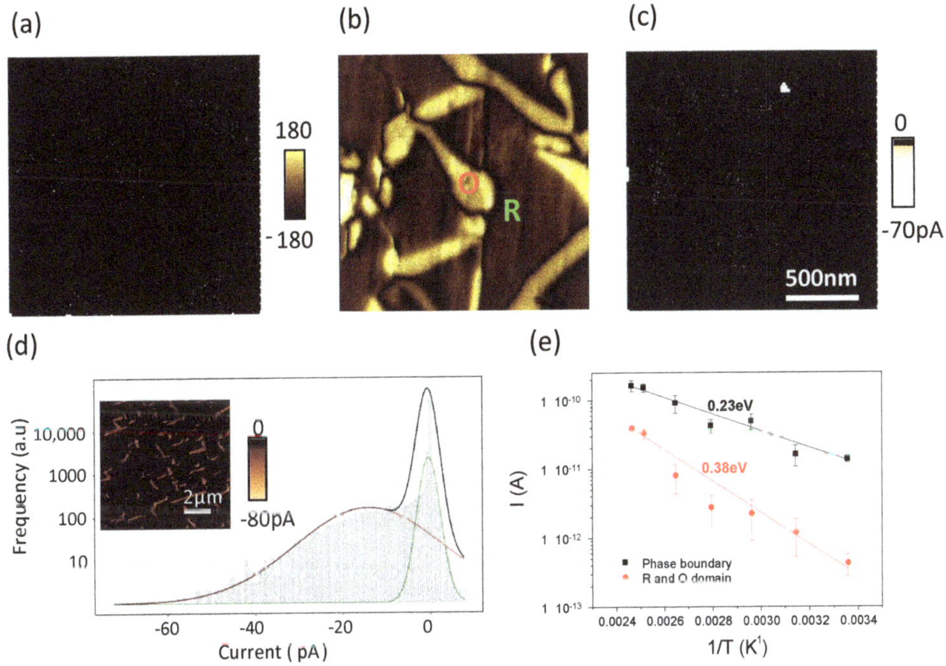

Figure 4. c-AFM based investigation of domain wall conduction, in this case for orthorhombic-rhombohedral phase boundaries (hybrid domain walls) in $BiFeO_3$. (**a**) PFM phase; (**b**) PFM amplitude image showing the two structural phases; (**c**) c-AFM revealing phase boundary electrical conductivity; (**d**) histogram analysis of conduction; (**e**) temperature-dependent c-AFM data showing thermally activated conduction mechanism.

Electronic properties of materials have been studied by scanning tunnelling microscopy and spectroscopy (STM), Kelvin probe force microscopy (KPFM), and scanning microwave impedance microscopy (sMIM). This includes domain walls, for example the variation of electronic band structure at individual walls [16]. Local changes of dielectric constant at structural domain walls have been studied in vanadium oxide by sMIM [28], which are especially pronounced for metal insulator transitions, but can also be used for variations in carrier density in semiconducting materials.

Among the SPM techniques available for the study of optical materials are near-field scanning optical microscopy (NSOM), and nanoscale infrared spectroscopy (nano-IR). They have been used to study strain induced variations of optical constants around domain walls [21], and the local variation of phonons around engineered domain walls in van der Waals heterostructure materials [20]. Various other SPM modes can also be combined with optical illumination to study the impact of light on topological defects, for example the displacement of domain walls in ferroelectrics [36–38] or optically driven vertex formation [39–41].

In addition to the imaging of various properties, SPM techniques can also be used to manipulate topological defects through the application of local stimuli, such as mechanical stress [42], electric and magnetic fields, and the study of topological defects in prototype device structures. One example is the demonstration and investigation of information storage in ferroelectric domain wall non-volatile memory cells [43,44] (Figure 5). Here the conduction paths of individual conductive domain walls can be followed, and polarization orientation can be studied simultaneously, providing information on polarization orientation across individual walls in memory device prototypes.

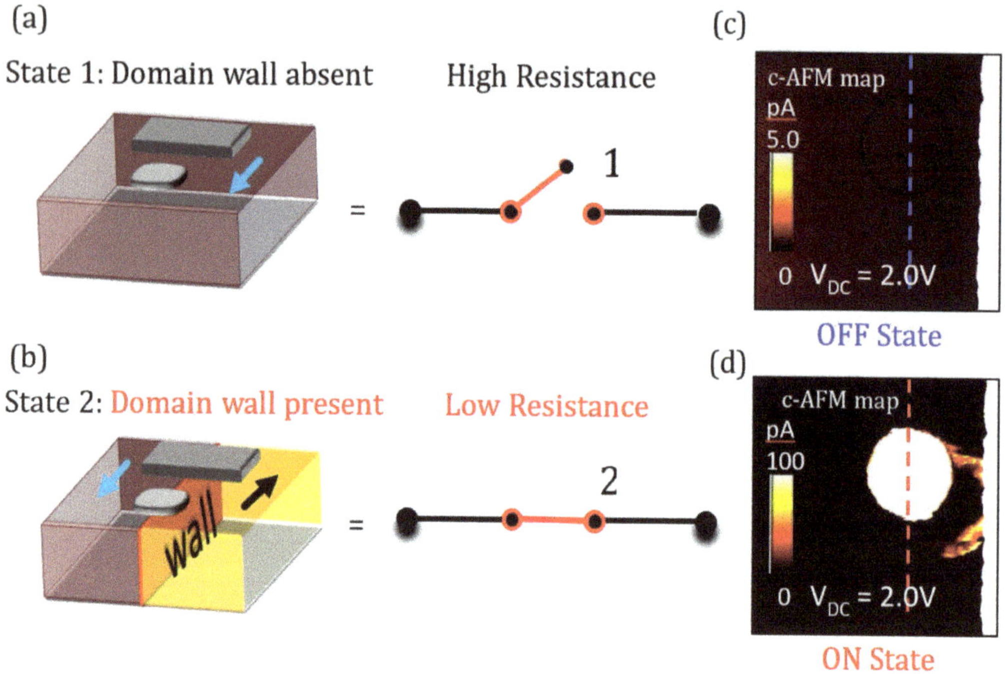

Figure 5. Ferroelectric domain wall memory. (**a**) high resistance state, domain wall is absent; (**b**) low resistance state, domain wall is present; (**c**) and (**d**) corresponding experimental realisations of (**a**) and (**b**), respectively, using c-AFM imaging of domain wall conduction path [43].

4. Concluding Remarks and Outlook

The study of materials surfaces and cross sections [45] with physical SPM probes enables the study of various properties associated with topological defects in solid state material systems, from fundamental material properties, to application relevant properties [46] and prototype devices [40,41]. The achieved lateral resolution is in many cases associated with the probe size (tip radius, contact area), which is sufficient to study mesoscopic nanoscale systems just above atomic resolution, or for measurements in UHV, at atomic resolution. It can also be used in correlative microscopy studies that involve other imaging and spectroscopy techniques, where SPM is used to localize features of interest or study corresponding properties.

Funding: This research received no external funding.

Institutional Review Board Statement: Not applicable.

Informed Consent Statement: Not applicable.

Data Availability Statement: Not applicable.

Conflicts of Interest: The author declares no conflict of interest.

References

1. Mermin, N.D. The topological theory of defects in ordered media. *Rev. Mod. Phys.* **1979**, *51*, 591. [CrossRef]
2. Přívratská, J.; Janovec, V. Pyromagnetic domain walls connecting antiferromagnetic non-ferroelastic magnetoelectric domains. *Ferroelectrics* **1997**, *204*, 321. [CrossRef]
3. Přívratská, J.; Janovec, V. Spontaneous polarization and/or magnetization in non-ferroelastic domain walls: Symmetry predictions. *Ferroelectrics* **1999**, *222*, 23. [CrossRef]
4. Catalan, G.; Seidel, J.; Ramesh, R.; Scott, J.F. Domain wall nanoelectronics. *Rev. Mod. Phys.* **2012**, *84*, 119. [CrossRef]
5. Parkin, S.S.P.; Hayashi, M.; Thomas, L. Magnetic Domain-Wall Racetrack Memory. *Science* **2008**, *320*, 190–194. [CrossRef]
6. Seidel, J. (Ed.) *Topological Structures in Ferroic Materials: Domain Walls, Vortices and Skyrmions*; Springer: Berlin, Germany, 2016; ISBN 978-3-319-25299-5. [CrossRef]
7. Seidel, J.; Vasudevan, R.K.; Nagarajan, V. Topological structures in multiferroics: Domain walls, skyrmions and vortices. *Adv. Electron. Mater.* **2016**, *2*, 1500292. [CrossRef]
8. Muhlbauer, S.; Binz, B.; Jonietz, F.; Pfleiderer, C.; Rosch, A.; Neubauer, A.; Georgii, R.; Boni, P. Skyrmion Lattice in a Chiral Magnet. *Science* **2009**, *323*, 915. [CrossRef]
9. Yu, X.Z.; Onose, Y.; Kanazawa, N.; Park, J.H.; Han, J.H.; Matsui, Y.; Nagaosa, N.; Tokura, Y. Real-space observation of a two-dimensional skyrmion crystal. *Nature* **2010**, *465*, 901. [CrossRef]
10. Balke, N.; Winchester, B.; Ren, W.; Chu, Y.H.; Morozovska, A.N.; Eliseev, E.A.; Huijben, M.; Vasudevan, R.K.; Maksymovych, P.; Britson, J.; et al. Enhanced electric conductivity at ferroelectric vortex cores in $BiFeO_3$. *Nat. Phys.* **2012**, *8*, 81–88. [CrossRef]
11. Göbel, B.; Mertig, I.; Tretiakov, O.A. Beyond skyrmions: Review and perspectives of alternative magnetic quasiparticles. *Phys. Rep.* **2021**, *895*, 1. [CrossRef]
12. Bert, V. Atomic Force Microscopy. In *NanoScience and Technology*; Springer: Berlin, Germany, 2019.
13. Giessibl, F.J. Advances in atomic force microscopy. *Rev. Mod. Phys.* **2003**, *75*, 949. [CrossRef]
14. Kalinin, S.V.; Rodriguez, B.J.; Kholkin, A.L. Piezoresponse Force Microscopy and Spectroscopy. In *Encyclopedia of Nanotechnology, Bhushan, B., Ed.*; Springer: Dordrecht, Netherlands, 2012. [CrossRef]
15. Gruverman, A.; Alexe, M.; Meier, D. Piezoresponse force microscopy and nanoferroic phenomena. *Nat. Commun.* **2019**, *10*, 1661. [CrossRef] [PubMed]
16. Romming, N.; Hanneken, C.; Menzel, M.; Bickel, J.E.; von Bergmannn, B.W.K.; Kubetzka, A.; Wiesendanger, R. Writing and Deleting Single Magnetic Skyrmions. *Science* **2013**, *341*, 636. [CrossRef] [PubMed]
17. Chiu, Y.-P.; Chen, Y.-T.; Huang, B.-C.; Shih, M.-C., Yang, J. C.; He, Q.; Liang, C.-W.; Seidel, J.; Chen, Y.-C.; Ramesh, R.; et al. Atomic-scale evolution of local electronic structure across multiferroic domain walls. *Adv. Mater.* **2011**, *23*, 1530. [CrossRef] [PubMed]
18. Turner, P.W.; McConville, J.P.V.; McCaftan, S.J.; Campbell, M.H.; Schaab, J.; McQuaid, R.G.P.; Kumar, A.; Gregg, J.M. Large Carrier Mobilities in $ErMnO_3$ Conducting Domain Walls Revealed by Quantitative Hall-Effect Measurements. *Nano Lett.* **2018**, *18*, 6381. [CrossRef] [PubMed]
19. Seidel, J.; Martin, L.W.; He, Q.; Zhan, Q.; Chu, Y.H.; Rother, A.; Hawkridge, M.; Maksymovych, P.; Kalinin, S.; Gemming, S.; et al. Conduction at domain walls in oxide multiferroics. *Nat. Mater.* **2009**, *8*, 229. [CrossRef]
20. Meier, D.; Seidel, J.; Cano, A.; Delaney, K.; Kumagai, Y.; Mostovoy, M.; Spaldin, N.A.; Ramesh, R.; Fiebig, M. Anisotropic conductance at improper ferroelectric domain walls. *Nat. Mater.* **2012**, *11*, 284. [CrossRef]
21. Moore, S.L., Ciccarino, C.J.; Halbertal, D.; McGilly, L.J.; Finney, N.R.; Yao, K.; Shao, Y.; Ni, G.; Sternbach, A.; Telford, E.J. Nanoscale lattice dynamics in hexagonal boron nitride moire superlattices. *Nat. Commun.* **2021**, *12*, 5741. [CrossRef]
22. Eng, L.M.; Guntherodt, H.J. Scanning force microscopy and near-field scanning optical microscopy of ferroelectric and ferroelastic domain walls. *Ferroelectrics* **2000**, *236*, 35. [CrossRef]
23. Seidel, J.; Eng, L.M. Shedding light on nanoscale ferroelectrics. *Curr. Appl. Phys.* **2014**, *14*, 1083. [CrossRef]
24. Milde, P.; Köhler, D.; Seidel, J.; Eng, L.M.; Bauer, A.; Chacon, A.; Kindervater, J.; Mühlbauer, S.; Pfleiderer, C.; Buhrandt, S.; et al. Unwinding of a skyrmion lattice by magnetic monopoles. *Science* **2013**, *340*, 1076. [CrossRef] [PubMed]
25. Kézsmárki, I.; Bordács, S.; Milde, P.; Neuber, E.; Eng, L.M.; White, J.S.; Rønnow, H.M.; Dewhurst, C.D.; Mochizuki, M.; Yanai, K.; et al. Néel-type skyrmion lattice with confined orientation in the polar magnetic semiconductor GaV4S8. *Nat. Mater.* **2015**, *14*, 1116. [CrossRef] [PubMed]
26. Persky, E.; Kalisky, B. Scanning SQUID view of oxide interfaces. *Adv. Mater.* **2018**, *30*, 1706653. [CrossRef] [PubMed]
27. Gross, I.; Akhtar, W.; Garcia, V.; Martinez, L.J.; Chouaieb, S.; Garcia, K.; Carretero, C.; Barthelemy, A.; Appel, P.; Maletinsky, P.; et al. Real-space imaging of non-collinear antiferromagnetic order with a single-spin magnetometer. *Nature* **2017**, *549*, 252. [CrossRef]
28. Rondin, L.; Tetienne, J.-P.; Hingant, T.; Roch, J.-F.; Maletinsky, P.; Jacques, V. Magnetometry with nitrogen-vacancy defects in diamond. *Rep. Prog. Phys.* **2014**, *77*, 056503. [CrossRef]
29. Tselev, A.; Meunier, V.; Strelcov, E.; Shelton, W.A.; Luk'yanchuk, I.; Jones, K.; Proksch, R.; Kolmakov, A.; Kalinin, S.V. Mesoscopic Metal-Insulator Transition at Ferroelastic Domain Walls in VO_2. *ACS Nano* **2010**, *4*, 4412. [CrossRef]

30. Ma, E.Y.; Cui, Y.-T.; Ueda, K.; Tang, S.; Chen, K.; Tamura, N.; Wu, P.M.; Fujioka, J.; Tokura, Y.; Shen, Z.-X. Mobile metallic domain walls in an all-in-all-out magnetic insulator. *Science* **2015**, *350*, 538. [CrossRef]
31. Hong, S.; Tong, S.; Park, W.I.; Hiranaga, Y.; Cho, Y.; Roelofs, A. Charge gradient microscopy. *Proc. Natl. Acad. Sci. USA* **2014**, *111*, 6566. [CrossRef]
32. Esfahani, E.N.; Liu, X.; Li, J. Imaging ferroelectric domains via charge gradient microscopy enhanced by principal component analysis. *J. Mater.* **2017**, *3*, 280. [CrossRef]
33. Salje, E.K.H. *Phase Transitions in Ferroelastic and Co-Elastic Crystals*; Cambridge University Press: Cambridge, UK, 1990.
34. Heo, Y.; Sharma, P.; Liu, Y.; Li, J.; Seidel, J. Mechanical probing of ferroelectrics at the nanoscale. *J. Mater. Chem. C* **2019**, *7*, 12441. [CrossRef]
35. Heo, Y.; Lee, J.H.; Xie, L.; Pan, X.; Yang, C.-H.; Seidel, J. Enhanced conductivity at orthorhombic-rhombohedral phase boundaries in $BiFeO_3$ thin films. *Nat. Publ. Group Asia Mater.* **2016**, *8*, e297. [CrossRef]
36. Vats, G.; Bai, Y.; Zhang, D.; Juuti, J.; Seidel, J. Optical control of ferroelectric domains: Nanoscale insight on macroscopic observations. *Adv. Opt. Mater.* **2019**, *7*, 1800858. [CrossRef]
37. Wang, J.; Liu, G.; Sando, D.; Nagarajan, V.; Seidel, J. Morphology-dependent photo-induced polarization recovery in ferroelectric thin films. *Appl. Phys. Lett.* **2017**, *111*, 092902. [CrossRef]
38. Vats, G.; Bai, Y.; Seidel, J. Opto-mechanical mapping of ferroelectric domains and the piezo-photovoltaic effect in Ba- and Ni-doped $(K_{0.5}Na_{0.5})NbO_3$. *Adv. Photonics Res.* **2021**, *2*, 2100050. [CrossRef]
39. Ravnik, J.; Askivskyi, I.; Gerasimenko, Y.; Diego, M.; Vodeb, J.; Kabanov, V.; Mihailovic, D.D. Strain-Induced Metastable Topological Networks in Laser-Fabricated TaS_2 Polytype Heterostructures for Nanoscale Devices. *ACS Appl. Nano Mater.* **2019**, *2*, 3743. [CrossRef]
40. Gerasimenko, Y.A.; Karpov, P.; Vaskivskyi, I.; Brazovskii, S.; Mihailovic, D. Intertwined Chiral Charge Orders and Topological Stabilization of the Light-Induced State of a Prototypical Transition Metal Dichalcogenide. *Npj Quantum Mater.* **2019**, *4*, 1. [CrossRef]
41. Karpov, P.; Brazovskii, S. Modeling of Networks and Globules of Charged Domain Walls Observed in Pump and Pulse Induced States. *Sci. Rep.-Uk* **2018**, *8*, 1. [CrossRef]
42. Vats, G.; Ravikant; Schönherr, P.; Kumar, A.; Seidel, J. Low-pressure mechanical switching of ferroelectric domains in $PbZr_{0.48}Ti_{0.52}O_3$. *Adv. Electron.* **2020**, *6*, 2000523.
43. Sharma, P.; Sando, D.; Zhang, P.; Cheng, X.; Prosandeev, S.; Bulanadi, R.; Prokhorenko, S.; Bellaiche, L.; Chen, L.-Q.; Nagarajan, V.; et al. Conformational domain wall switch. *Adv. Funct. Mater.* **2019**, *29*, 1807523. [CrossRef]
44. Sharma, P.; Moise, T.S.; Colombo, L.; Seidel, J. Roadmap for ferroelectric domain wall nanoelectronics. *Adv. Funct. Mater.* **2022**, *32*, 2110263. [CrossRef]
45. Ji, F.; Yao, Y.; Xin, T.; Seidel, J. A comprehensive FIB lift-out sample preparation method for scanning probe microscopy. *Nanomanuf. Metrol.* **2022**, *5*, 67–79. [CrossRef]
46. Zhang, D.; Sando, D.; Sharma, P.; Cheng, X.; Ji, F.; Govinden, V.; Weyland, M.; Nagarajan, V.; Seidel, J. Superior polarization retention through engineered domain wall pinning. *Nat. Commun.* **2020**, *11*, 349. [CrossRef] [PubMed]

symmetry

MDPI

Article

Assessing Bound States in a One-Dimensional Topological Superconductor: Majorana versus Tamm

Lucia Vigliotti [1], Fabio Cavaliere [1,2], Matteo Carrega [2] and Niccolò Traverso Ziani [1,2,*]

1 Dipartimento di Fisica Università di Genova, Via Dodecaneso 33, 16146 Genova, Italy; lucia.vigliotti@edu.unige.it (L.V.); fabio.cavaliere@unige.it (F.C.)
2 CNR-Spin, Via Dodecaneso 33, 16146 Genova, Italy; matteo.carrega@spin.cnr.it
* Correspondence: traversoziani@fisica.unige.it

Abstract: Majorana bound states in topological superconductors have attracted intense research activity in view of applications in topological quantum computation. However, they are not the only example of topological bound states that can occur in such systems. Here, we study a model in which both Majorana and Tamm bound states compete. We show both numerically and analytically that, surprisingly, the Tamm state remains partially localized even when the spectrum becomes gapless. Despite this fact, we demonstrate that the Majorana polarization shows a clear transition between the two regimes.

Keywords: Majorana fermions; topological superconductors; Majorana polarization

1. Introduction

Non-Abelian quasiparticles have attracted intense theoretical and experimental research activities in the last several years, in view of their potential applications in topologically protected quantum computation [1]. Several proposals for the experimental realization or detection of such exotic states of matter in interacting systems have hence been put forward [2–13]. Despite steady progress, the need to enforce the coexistence of superconductivity and electronic interactions makes their experimental realization challenging [14–16]. In the absence of interactions, the possibility to engineer non-Abelian Majorana zero modes [17] in spinless p-wave superconductors has been predicted [18,19]. The unambiguous experimental realization of such states is again far from straightforward, due to the fact that spinless p-wave superconductivity is rare in nature. However, it has been realized that this kind of pairing could be achieved by means of proximity-induced superconductivity on semiconducting structures with pronounced spin-dependent properties. One of the first example of a system that, once proximitized, can host Majorana bound states is represented by topological insulators [20–22], where strong spin–orbit coupling gives rise to inverted bands behavior. However, it additionally requires the presence of magnetic barriers to allow for Majorana bound state formation. The difficulty in implanting magnetic barriers has slowed down the development of such a platform. Recently, the possibility to employ quantum point contacts [23] instead of magnetic barriers has been considered, providing a promising alternative for topological insulator-based Majorana platforms [24,25]. Several other platforms have been proposed, ranging from ferromagnetic atomic chains with superconducting pairing [26], planar Josephson junctions in 2D semiconductors [27–29], and the more popular setup based on spin–orbit-coupled quantum wire with proximity superconductivity and applied magnetic fields [30,31].

In the last case, several experiments have reported signatures compatible with the presence of Majorana zero modes [32]. However, a clear evidence of the quantization of the zero bias peak, which is one of the predicted signatures ascribed to the presence of Majorana zero modes, or other unambiguous fingerprints of their formation are still lacking to date. Indeed, an additional challenge in the field is posed by the fact that Majorana zero modes behave, in many terms, similarly to more conventional Andreev bound states, which do not exhibit non-Abelian statistics, and are thus less useful in quantum computation [33–50]. Indeed, on one hand, it is still possible to distinguish between trivial Andreev bound states that emerge at finite chemical potential in clean systems and Majorana bound states due to the fact that the phase containing Majorana fermions and the one with the trivial bound states are separated by a region with extended states only [51]. On the other hand, the distinction becomes more subtle in the presence of disorder, spatial variations of the confinement potentials, or whenever the formation of quantum dots [52] takes place. A detailed analysis of the scenarios in which Majorana fermions and other bound states coexist is hence in order.

In this work, we concentrate on this aspect by studying a simple model where different bound states may coexist. To this end, we consider a one-dimensional finite size spinless p-wave superconductor in the presence of a spatially periodic modulation of the local chemical potential. The system is numerically investigated by means of extensive exact diagonalization. We find that the bulk of the system is always gapped at the chemical potential, except for a gapless point that marks the boundary between two phases dominated either by the periodic local potential (A) or the superconducting one (B). In the A phase, either a bound state is localized at one end of the system or no boundary states are present, depending on the phase of the modulated potential. The possible boundary state is adiabatically connected with the system in the absence of superconductivity and hence is, qualitatively speaking, a Tamm state [53–55]. In the B phase, one recovers the more conventional Majorana scenario, with the zero mode split into two Majorana modes at both ends of the system. Surprisingly, at the transition between A and B, the Tamm state remains partially localized. This behavior poses the question if the Majorana fermions in the B phase are still topological in nature or if the reminiscence of the Tamm state eventually spoils their properties.

In the second part of the work, we answer this question, deriving and analyzing a linearized low-energy model that can be analytically solved. This allows us to interpret and discuss the physics observed in the first part in terms of Goldstone–Wilczek charges [54,56–63].

Finally, to elucidate the interplay between the different kinds of bound states, and to possibly discriminate between them, we evaluate the Majorana polarization [64–66] associated to the boundary states. This quantity is one of the standard tools introduced to characterize topological superconductors. This allows us to conclude that phase B is indeed characterized by the presence of Majorana fermions.

In more details, the article is divided as follows. In Section 2, we analyze the results for the p-wave superconductor model. In Section 3, we derive the exactly solvable model and discuss it. Section 4 contains the calculation of the Majorana polarization. Our conclusions are finally drawn in Section 5.

2. The Quadratic Model

2.1. Hamiltonian

The system we inspect is a finite-size, spinless one-dimensional p-wave superconductor in the presence of an additional periodic potential [67]. More specifically, the Hamiltonian, defined on a segment of length L, is

$$H = \frac{1}{2}\int_0^L \Psi^\dagger(x)\mathcal{H}(x)\Psi(x)dx, \tag{1}$$

with the Bogoliubov–de Gennes (BdG) Hamiltonian density $\mathcal{H}(x)$ given by ($\hbar = 1$)

$$\mathcal{H}(x) = \left(\frac{-\partial_x^2}{2m} - \mu_0 - \mu_1(x)\right)\tau_3 + (-i)\frac{\Delta}{k_F}\partial_x\tau_1, \tag{2}$$

and $\Psi(x) = (\psi(x), \psi^{\dagger}(x))^T$, where $\psi(x)$ is a fermionic annihilation field. We impose open boundary conditions, implying $\Psi(0) = \Psi(L) = 0$. In the Hamiltonian, m is the effective mass, μ_0 sets the filling with (Note that, in view of the finite size of the system, we also impose $k_F = n_F\pi/L$, with n_F a positive integer.) $k_F = \sqrt{2m\mu_0}$, Δ the strength of the p-wave superconducting pairing and τ_i $(i = 1, 2, 3)$ are the Pauli matrices. For future use, we also introduce τ_0 as the 2X2 identity matrix. Finally, we set

$$\mu_1(x) = -B\cos(2k_F x + \varphi), \tag{3}$$

with the phase φ kept as a free parameter. Here, $\mu_1(x)$ is a periodic potential that can emerge due to external gates [68] or via a coupling to the phonons [69]. The parameter B parametrizes the strength of the term, while the phase φ, which is clearly an irrelevant parameter when periodic boundary conditions are imposed, becomes essential in the case of open boundary conditions. The manipulation of the phase φ can be envisioned if $\mu_1(x)$ is due to external finger gates [68]. In this case, the spatial variations of the potential can be fully manipulated.

The energy spectrum and the wavefunctions are evaluated via a numerical exact diagonalization procedure, which consists of expanding the eigenfunctions $\psi_\nu(x)$ of the Hamiltonian on the basis of the states of a free particle in a infinite 1D box of length L

$$\psi_\nu(x) = \sqrt{\frac{2}{L}}\sum_n \sin\left(\frac{n\pi x}{L}\right) c_n(\nu) \tag{4}$$

and diagonalizing the associated matrix of the Hamiltonian on this basis. This allows us to obtain the energy spectrum E_ν and the weights $c_n(\nu)$ of the eigenfunctions of the problem. It is here worth mentioning that since we adopt the Bogoliubov–de Gennes formalism, we artificially double the spectrum by introducing a redundant chiral symmetry, and hence only half of the eigenfunctions of the Hamiltonian matrix have physical meaning. Indeed, by only taking the positive energy eigenstates, one recovers the correct excitation spectrum. The diagonalization is performed in Mathematica (TM) and up to 450 box states per BdG sector are employed depending on the values of B and Δ, to ensure numerical convergence with a relative error $\delta \lesssim 10^{-3}$ on the energy spectrum.

2.2. Tamm States

We begin our analysis by considering the case $\Delta = 0$.

Figure 1 shows the energy spectrum for different values of the strength B and different phases φ. As a general feature, we observe that, for $B \neq 0$, the periodic potential opens a gap at zero energy—where the chemical potential is set. Secondary gaps can occur when the amplitude B becomes large, $B \gtrsim \mu_0$. This additional gap, visible in Figure 1a for $\nu \sim 150$, opens, having in mind periodic boundary conditions and a folded scheme for taking into account the periodic perturbation, at $k = 0$. For the same magnitudes of B, an additional feature appears for small ν (see the blue and red curves). This is due to the change in nature of the lowest energy states, which, for $B > \mu_0$, are dominated by the periodic potential. Comparing Figure 1a,b, one can observe that for almost all the states, the overall shape of the energy spectrum depends weakly on the phase φ, being almost insensitive to it for $E \gtrsim \mu_0$. However, inspecting the case $\varphi = \pi/2$, one can observe that a zero-energy mode occurs within the gap, which suggests that the low-energy sector of the spectrum may exhibit a more pronounced dependence on φ.

To check this fact, Figure 2a shows the behavior of the first eight eigenstates with $E_\nu \geq 0$ as a function of φ for the representative case of $B = \mu_0/2$. As one can clearly see, the lowest energy state exhibits a quite dramatic dependence on the phase, becoming a zero mode for $\varphi = (2p+1)\pi/2$ with p an integer. The wavefunction of such a zero-energy state is shown for $\varphi = \pi/2$ in Figure 2b and it corresponds to a localized state, which, in this context, is usually referred to as a Tamm state. Such a state is respectively localized either at the right ($\varphi = \pi/2 + 2\pi p$) or at the left ($\varphi = 3\pi/2 + 2\pi p$) edge of the system.

Such a Tamm state is starkly different from the delocalized states obtained at $\varphi = \pi p$, which extend over the entire length of the system; see the representative example reported in Figure 2c for the case $\varphi = 0$. In general, for $\varphi \neq \pi p$, a bound state within the gap always occurs, possibly with a nonzero energy and a larger localization length.

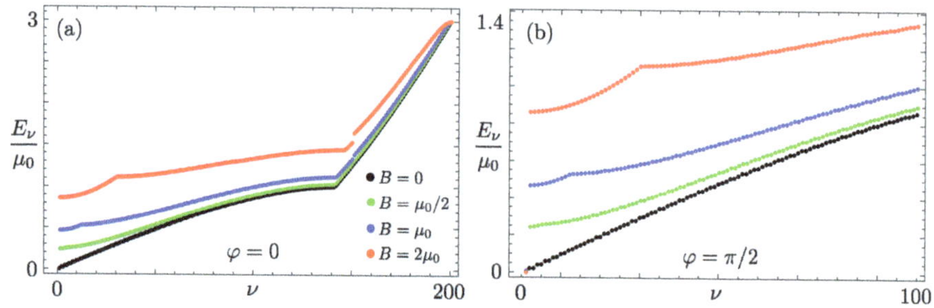

Figure 1. Plots of the energy spectrum for different values of the amplitude of the periodic potential: $B = 0$ (black); $B = \mu_0/2$ (green); $B = \mu_0$ (blue); $B = 2\mu_0$ (red). The case $\varphi = 0$ is shown in panel (**a**) for the first 200 eigenstates, while $\varphi = \pi/2$ in panel (**b**) for the first 100 eigenstates.

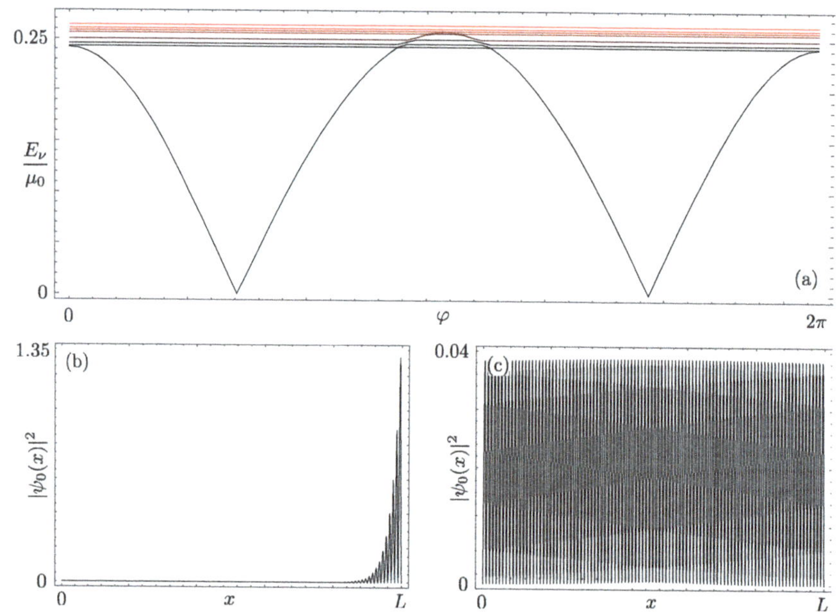

Figure 2. Plot of E_ν/μ_0 for $0 \leq \nu \leq 7$ (from black to red) as a function of φ (panel **a**) and probability density of the $\nu = 0$ state $|\psi_0(x)|^2$ (units L^{-1}) as a function of x for $\varphi = \pi/2$ (panel **b**) and $\varphi = 0$ (panel **c**). In all panels, the amplitude of the periodic potential is set at $B = \mu_0/2$.

2.3. Majorana States

The regime $B = 0$ with $\Delta \neq 0$ is more known since it represents the paradigmatic model that hosts Majorana bound states: even in this case a gap opens around the chemical potential and a fermionic zero-energy mode appears in the spectrum.

The BdG wave function corresponding to this fermionic state is significantly nonzero close to both ends of the segment (see Figure 3). Formally, one can interpret this nonlocal fermionic state as two local Majorana states, each one localized close to one end only. This is obviously meaningful only when the localization length is smaller than the length of the system.

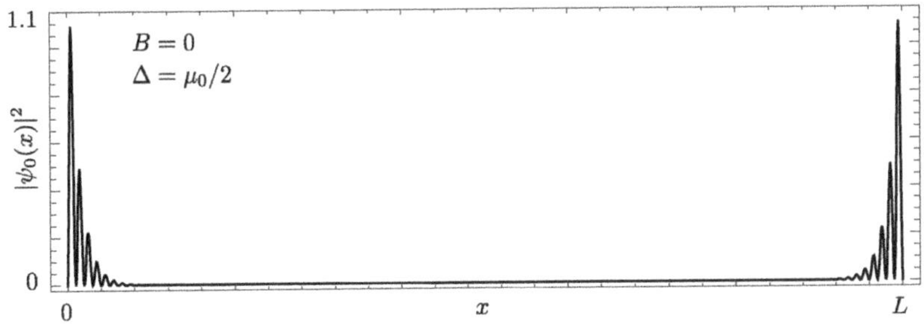

Figure 3. Probability density of the zero-energy Majorana state $|\psi_0(x)|^2$ (units L^{-1}) as a function of x for $B = 0$ and $\Delta = \mu_0/2$.

2.4. Competition between Tamm and Majorana States

Let us now address the case when both B and Δ are nonzero and thus the two gaps induced by the oscillating potential and by the p-wave pairing compete. We are particularly interested into the interplay between the two (qualitatively different) localized Tamm and Majorana states, and thus we set from now on $\varphi = \pi/2$ to achieve the most dramatic effects. Indeed, $\varphi = \pi/2$ corresponds to a maximally localized Tamm state. As a general feature, we find that in this case, for any value of B, Δ a zero-energy mode occurs.

Figure 4a–e shows the probability amplitude $|\psi_0(x)|^2$ of this zero-energy mode for $B = \mu_0/4$ and increasing values of the superconducting pairing. As Δ is increased from zero, the localized state broadens (see panel (b)). However, when $\Delta = B$ (see panel (c)), a peculiar phenomenon occurs: the former Tamm state gets sharper once again while the wavefunction in the rest of the system becomes nonzero and with a flat envelope. When $\Delta > B$ (see panel (d)), a second localized state develops. Finally, when $\Delta \gg B$, the wavefunction tends to recover a symmetric shape localized at both ends of the system.

To better understand what is going on, we study the spectral weight $w(n, E) = |c_n[\nu(E)]|^2$, shown as a density plot as a function of the box state index n and the energy E in Figure 5 for energies around $E = 0$. (Note that due to the finite size of the system, the spectrum is always discrete and a binning of the energies has been performed to produce the plots.) In general, this tool allows us to get a glimpse of the structure of the energy spectrum of the system. Indeed, panel (a) shows, as a warm-up, the case $B = \Delta = 0$. As one can see, the spectral weight is sharply peaked on one box state per energy value and the yellow traces faithfully reproduce the energy spectra of the particle ($E > 0$) and hole ($E < 0$) sectors of the BdG picture for the Hamiltonian $H = \frac{p^2}{2m} - \mu_0$. Panel (b) shows the situation for $B = \mu_0/4$ and $\Delta = 0$, when a Tamm state occurs. The energy gap is clearly visible and the blurry feature at $E = 0$ is the spectral representation of the Tamm state, whose amplitude is smeared across many delocalized box states in order to produce a state localized in space. The situation is somewhat similar for $\Delta > B$—see panel (d)—again consistent with the picture of a wavefunction with sharp peaks at the system ends. However, the situation for $B = \Delta$, shown in panel (c), is peculiar. At this transition point, the spectrum is gapless and exhibits two linear modes around $E = 0$. Thus, the zero-energy mode is embedded into a gapless spectrum, and yet it still represents a bound state.

To summarize our findings so far, even though a gap closing is present at the level of the bulk spectrum, bound states can exist for every value of B and Δ. However, assessing the nature of such bound states (Tamm- or Majorana-like) is not a trivial affair in the context of the full quadratic model, which can only be solved numerically. A notable exception is represented by the point $\varphi = 0$. In that case, indeed, no Tamm state is present and a bound state is necessarily a Majorana state. However, the general case needs further inspection. In order to tackle this problem and provide an answer, we now proceed with development of an analytically solvable low-energy model that is able to capture all of the features described so far.

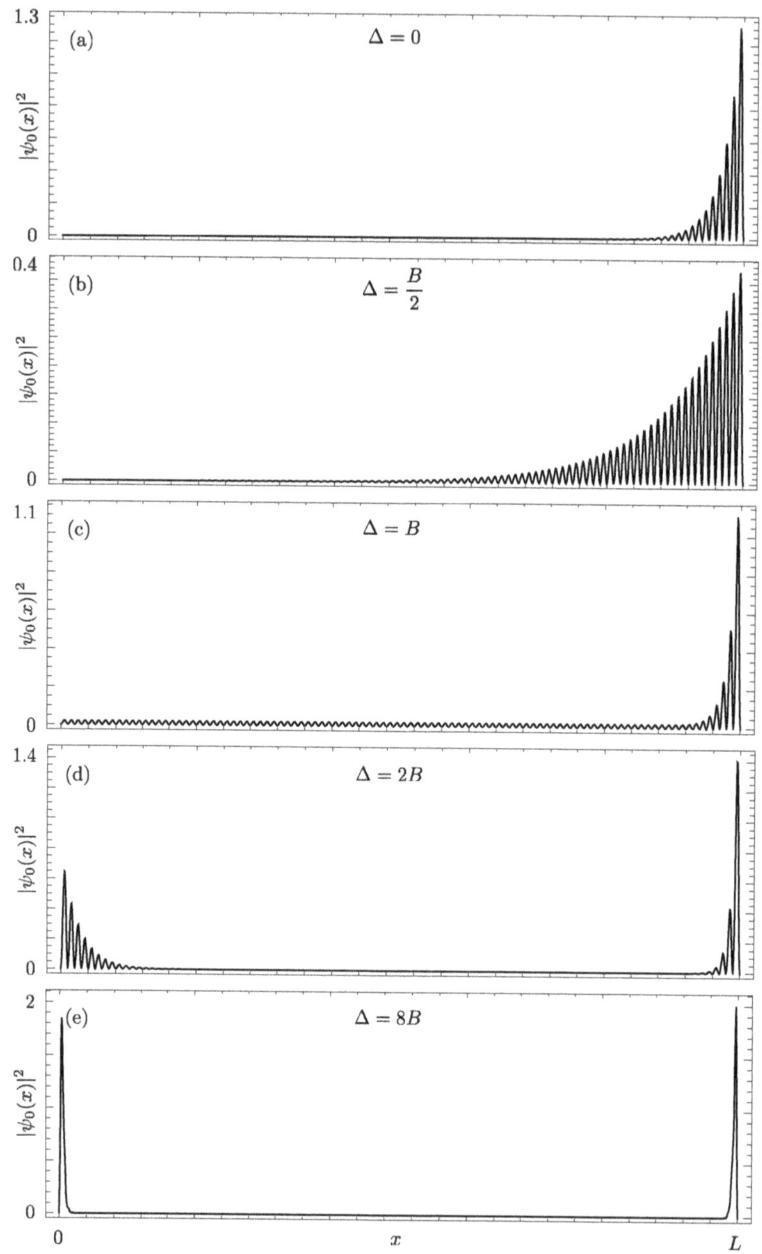

Figure 4. Probability density of the zero-energy state $|\psi_0(x)|^2$ (units L^{-1}) as a function of x for different values of the p-wave pairing strength: $\Delta = 0$ (**a**); $\Delta = B/2$ (**b**); $\Delta = B$ (**c**); $\Delta = 2B$ (**d**); $\Delta = 8B$ (**e**). In all panels, $B = \mu_0/4$ and $\varphi = \pi/2$.

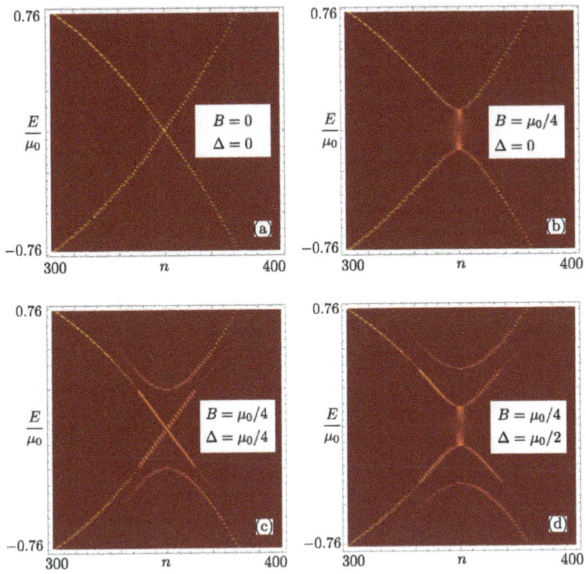

Figure 5. Density plot of the spectral weight $w(n, E)$ (see text) as a function of the index n labeling the eigenstates of a particle in a box of length L and energy E (units μ_0) for different values of B and Δ: $B = \Delta = 0$ (**a**); $B = \mu_0/4, \Delta = 0$ (**b**); $B = \Delta = \mu_0/4$ (**c**); $B = \mu_0/4, \Delta = 2B$ (**d**).

3. Effective Low-Energy Model

3.1. Linearization

To develop an effective, low-energy model valid around the chemical potential, we consider a linear dispersion relation instead of the quadratic one. This is meaningful as long as μ_0 is the largest energy scale involved [70–72]. In detail, we approximate the fermionic operator as [73–76]

$$\psi(x) \simeq e^{ik_F x}\psi_R(x) + e^{-ik_F x}\psi_L(x), \tag{5}$$

with $\psi_{R,L}(x)$ fermionic operators for right- or left-moving electrons. In particular, using the particle-in-a-box states, one can write

$$\psi_R(x) = \frac{-i}{\sqrt{2L}} \sum_{n=-\infty}^{\infty} e^{in\pi x/L} d_{n+n_F}, \tag{6}$$

and $\psi_L(x) = -\psi_R(-x)$ ensuring, together with the $2L$ periodicity of the fields, that open boundary conditions $\psi(0) = \psi(L) = 0$ are satisfied. Here, the fermionic operators d_n are associated to the nth eigenstate of a particle in a box. Note that, as an approximation, the sum is here extended to *all* integers and, consequently, a slight notation abuse has been adopted, having defined the operators d_n for all the integer values n. This approximation scheme is always performed when discussing the bosonization procedure, and is discussed, for example, in [77].

The kinetic energy is then approximated with K_L

$$K_L = v_F \sum_n \frac{n\pi}{L} d_{n+n_F} = \int_0^L \left[\psi_R^\dagger(x)(-iv_F\partial_x)\psi_R(x) + \psi_L^\dagger(x)(iv_F\partial_x)\psi_L(x)\right] dx, \tag{7}$$

with $v_F = (\pi n_F)/(mL)$ representing the Fermi velocity.

In order to write the full Hamiltonian within the linear approximation, it is useful to introduce the enlarged BdG spinor

$$\Psi_A(x) = (\psi_R(x), \psi_L(x), \psi_L^\dagger(x), -\psi_R^\dagger(x))^T. \tag{8}$$

By imposing that μ_0 is the largest energy scale (we indeed neglect terms proportional to $1/L$ with respect to terms proportional to k_F), we finally get the approximate form for the Hamiltonian

$$H \simeq \frac{1}{2}\int_0^L \Psi_A^\dagger(x)\mathcal{H}_A(x)\Psi_A(x)dx, \tag{9}$$

with

$$\mathcal{H}_A(x) = -iv_F\partial_x\tau_3 \otimes \tau_3 + B\tau_0 \otimes (\cos(\phi)\tau_1 + \sin(\phi)\tau_2) + \Delta\tau_1 \otimes \tau_0. \tag{10}$$

In order to interpret the boundary conditions and the peculiar effects that they may have on the bound states, it is useful to think in the following way. Consider a pair of chiral fermions $\chi_R(x)$ and $\chi_L(x)$, independent from each other and organized in the BdG spinor $X(x) = (\chi_R(x), \chi_L(x), \chi_L^\dagger(x), -\chi_R^\dagger(x))^T$, defined on the whole real axis, and with Hamiltonian

$$H_\chi = \frac{1}{2}\int_{-\infty}^{+\infty} dx\, X^\dagger(x)[\mathcal{H}_A(x) + M(x)]X(x), \tag{11}$$

where we have introduced a localized backscattering potential

$$M(x) = m_0[\delta(x) - \delta(x-L)]\tau_2, \tag{12}$$

with m_0 the backscattering strength. It can be easily shown in the limit $m_0/v_F \to \infty$ that the wavefunctions in the region $0 < x < L$ become disconnected and acquire the boundary conditions $\chi_L(0) = -\chi_R(0)$, $\chi_L(L) = -\chi_R(-L)$ [78–80]. The Schr'odinger problem associated to H_χ, when considered for $0 < x < L$ and when $m_0/v_F \to \infty$, is hence perfectly equivalent to the one related to H in Equation (9).

The advantage in resorting to H_χ is conceptual, since it allows to give a simple interpretation to the results described in the previous section. Indeed, from a physical point of view, H_χ is equivalent to the Hamiltonian of a quantum spin Hall liquid gapped by both superconductivity, with gap Δ, and a magnetic mass pointing in the direction given by ϕ in the presence of two strong backscattering centers with opposite magnetization localized at $x = 0, L$.

3.2. *Qualitative Interpretation of the Results*

The first insight into the results comes from the computation of the spectrum as obtained by neglecting the presence of boundaries. If we consider the Hamiltonian H_∞ given by

$$H_\infty = \frac{1}{2}\int_{-\infty}^{\infty} \Psi_A^\dagger(x)\mathcal{H}_A(x)\Psi_A(x)dx, \tag{13}$$

where the operators are now being defined, with a slight abuse of notation, on the whole real axis, one can obtain the energy spectrum. The spectrum is promptly found to be $\epsilon_j(k) = \pm\sqrt{v_F^2k^2 + \Delta^2 + B^2 + (-1)^j 2\Delta B}$ with $j \in \{1,2\}$ and is shown in Figure 6.

As can be seen, the spectrum is always gapped for all $B > 0$, $\Delta > 0$ except for $B = \Delta$—see panel (b)—where $\epsilon_1(k)$ is gapless. The phase space point $B = \Delta$ marks a quantum phase transition where the gap changes from magnetic ($B > \Delta$) to superconducting ($B < \Delta$). However, it is crucial to notice the presence of a secondary gap of magnitude $B + \Delta$ in $\epsilon_2(k)$, which is always present for every $B > 0$, $\Delta > 0$. Notice also the qualitative similarity between these spectra and the spectral features found in the quadratic model substantiating the validity of the linear approximation at low energies, compared to μ_0.

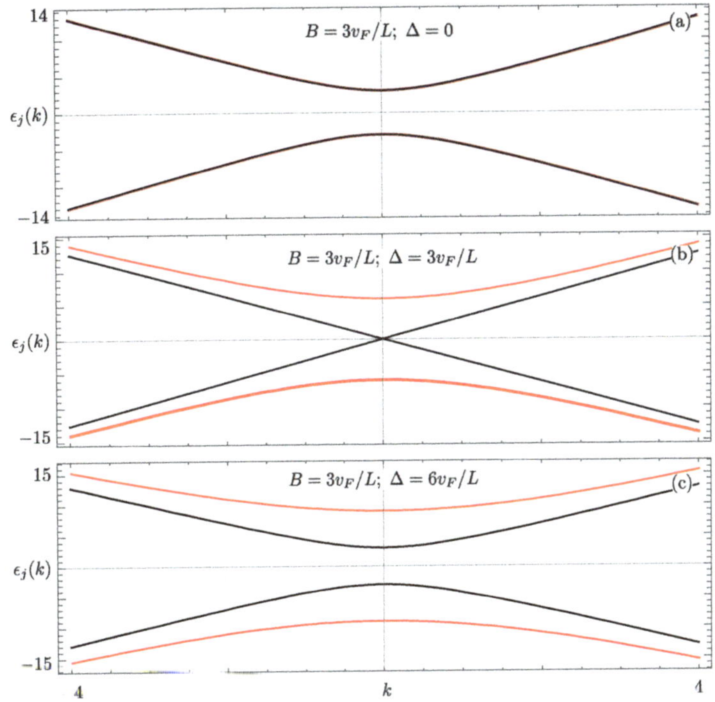

Figure 6. Plot of $\epsilon_j(k)$ (units v_F/L) as a function of k (units π/L) for: $B = 3v_F/L$, $\Delta = 0$ (**a**); $B = 3v_F/L$, $\Delta = 3v_F/L$ (**b**); $B = 3v_F/L$, $\Delta = 6v_F/L$ (**c**). In all panels, red lines represent $\epsilon_2(k)$; black lines correspond to $\epsilon_1(k)$.

Turning to the inspection of bound states, it is crucial to observe that the addition of a confining term proportional to m_0 produces a magnetic gap in the spectrum. It is well known [81] that in the superconducting dominated regime ($B < \Delta$), Majorana bound states appear at the boundaries of the system. However, in the opposite case, when $B > \Delta$ the situation is more subtle, as this model would implement the physics of so-called Goldstone–Wilczek fractional solitons [62]. Since this regime is adiabatically connected to $\Delta = 0$, in order to build some physical intuition, one can refer to the case in which superconductivity is absent. The system is then equivalent to a quantum spin Hall liquid, in the absence of superconductivity, gapped by two magnets of equal strength at $x = 0, L$ and characterized by different magnetization angles: it hence hosts bound states. Assuming the decay length of these bound states to be shorter than the system size L, the two bound states are essentially independent, and thus it is sufficient to concentrate on a single boundary to capture the relevant physics.

3.3. *Linear Model with One Boundary and* $\Delta = 0$

We thus inspect a simpler linear model in the regime $\Delta = 0$, $L \to \infty$ considering the bound state localized around $x = 0$. Its wavefunction $\psi_m(x)$, that since superconductivity is now absent ($\Delta = 0$), has only the two right- and left-moving components, can be easily obtained and reads

$$\psi_m(x) = \sqrt{\frac{m}{2v_F}\sin(\phi)}(-1,1)^T e^{\frac{m}{v_F}\sin(\phi)x}. \tag{14}$$

The solution is only acceptable (normalizable) for $\pi < \phi < 2\pi$. The corresponding energy is $\epsilon_0 = -m\cos(\phi)$. Note that for $\phi = 3\pi/2$ the state, mimicking the Tamm state of the quadratic model, is at zero energy. The scenario just derived perfectly agrees with the behavior of the left boundary of the quadratic model, although in that case the wavefunction in Equation (14) by means of Equation (5) has one component only, and is rebuilt from the condition in Equation (5). Moreover, in the quadratic model, the unphysical particle–hole symmetry of the Bogoliubov–de Gennes equation is present while here the redundancy [82,83] is not needed since superconductivity is not present. Hence, we have a single bound state instead of two at opposite energies.

The addition of a second physical boundary at $x = L$ produces effects that can be now easily understood by leveraging on the single boundary case just discussed. Indeed, the bound state at the boundary at $x = L$ is present when $\sin(\phi) > 0$—that is, for $0 < \phi < \pi$—when the bound state is absent at $x = 0$. Such exponentially increasing solution is admissible due to the added barrier at $x = L$—which prevents the divergence of the wavefunction—and its decay through the dot. This again is perfectly compatible with the results of the quadratic model discussed in Section 2.

A final comment for this section is the behavior at $\phi = 0, \pi$. For those values of the angle, the energy of the bound state reaches the bulk bands and the bound state becomes delocalized, and hence, effectively, disappears.

In the following, we will concentrate on the case in which the bound state is maximally localized around $x = 0$—that is, for $\phi = 3\pi/2$.

3.4. The Wavefunction of the Bound State for $\varphi = 3\pi/2$

A further interesting behavior of the bound state found within the quadratic model is the fact that, when present, it does not become completely delocalized at the quantum phase transition $B = \Delta$. To confirm that this fact occurs also within the linearized model, we here report the BdG expression of the wavefunction $\psi_0(x)$ of the bound state characterizing the linearized model in Equation (9), together with the condition in Equation (5), for $\phi = 3\pi/2$ and arbitrary $B > 0$, $\Delta > 0$. We find

$$\psi_0(x) = (u(x), v(x))^T, \tag{15}$$

where

$$\begin{aligned}
u(x) =& e^{ik_F x}\chi_0(x) - e^{-ik_F x}\chi_0(-x),\\
v(x) =& e^{ik_F x}\xi_0(x) - e^{-ik_F x}\xi_0(-x),
\end{aligned}$$

$$\begin{aligned}
\chi_0(x) =& \left[\left(\frac{\Delta - B}{\Delta + B}\frac{\sinh[L(\Delta+B)/v_F]}{\sinh[L(\Delta-B)/v_F]}\frac{\Delta+B}{4v_F}\frac{e^{-2\Delta L/v_F}}{1-e^{-2(\Delta+B)L/v_F}}\right)^{1/2} e^{(\Delta-B)x/v_F} + \right.\\
&\left. + \left(\frac{\Delta+B}{4v_F}\frac{1}{1-e^{-2(\Delta+B)L/v_F}}\right)^{1/2} e^{-(\Delta+B)x/v_F}\right]\Theta(x) +\\
&+ \left[\left(\frac{\Delta - B}{\Delta + B}\frac{\sinh[L(\Delta+B)/v_F]}{\sinh[L(\Delta-B)/v_F]}\frac{\Delta+B}{4v_F}\frac{e^{-2\Delta L/v_F}}{1-e^{-2(\Delta+B)L/v_F}}\right)^{1/2} e^{-(\Delta-B)x/v_F} + \right.\\
&\left. + \left(\frac{\Delta+B}{4v_F}\frac{1}{1-e^{-2(\Delta+B)L/v_F}}\right)^{1/2} e^{(\Delta+B)x/v_F}\right]\Theta(-x)\\
\xi_0(x) =& \left[\left(\frac{\Delta - B}{\Delta + B}\frac{\sinh[L(\Delta+B)/v_F]}{\sinh[L(\Delta-B)/v_F]}\frac{\Delta+B}{4v_F}\frac{e^{-2\Delta L/v_F}}{1-e^{-2(\Delta+B)L/v_F}}\right)^{1/2} e^{(\Delta-B)x/v_F} + \right.\\
&\left. - \left(\frac{\Delta+B}{4v_F}\frac{1}{1-e^{-2(\Delta+B)L/v_F}}\right)^{1/2} e^{-(\Delta+B)x/v_F}\right]\Theta(x) +\\
&+ \left[\left(\frac{\Delta - B}{\Delta + B}\frac{\sinh[L(\Delta+B)/v_F]}{\sinh[L(\Delta-B)/v_F]}\frac{\Delta+B}{4v_F}\frac{e^{-2\Delta L/v_F}}{1-e^{-2(\Delta+B)L/v_F}}\right)^{1/2} e^{-(\Delta-B)x/v_F} + \right.\\
&\left. - \left(\frac{\Delta+B}{4v_F}\frac{1}{1-e^{-2(\Delta+B)L/v_F}}\right)^{1/2} e^{(\Delta+B)x/v_F}\right]\Theta(-x).
\end{aligned}$$

Both the particle and the hole components $u(x)$ and $v(x)$ are characterized by a profile that is peaked around the edge, even at the transition point $B = \Delta$.

This case is shown in Figure 7, where a plot of $|u(x)|^2$ and $|v(x)|^2$ is reported for $B = \Delta$: one can clearly notice that, although the wavefunction is sharply localized at $x = 0$, it does not completely decay exponentially to zero at the transition point. This fact is in excellent agreement with the results obtained in the full quadratic model. As apparent from the analytical form of the wavefunction, this fact is due to the presence of the $B + \Delta$ energy scale.

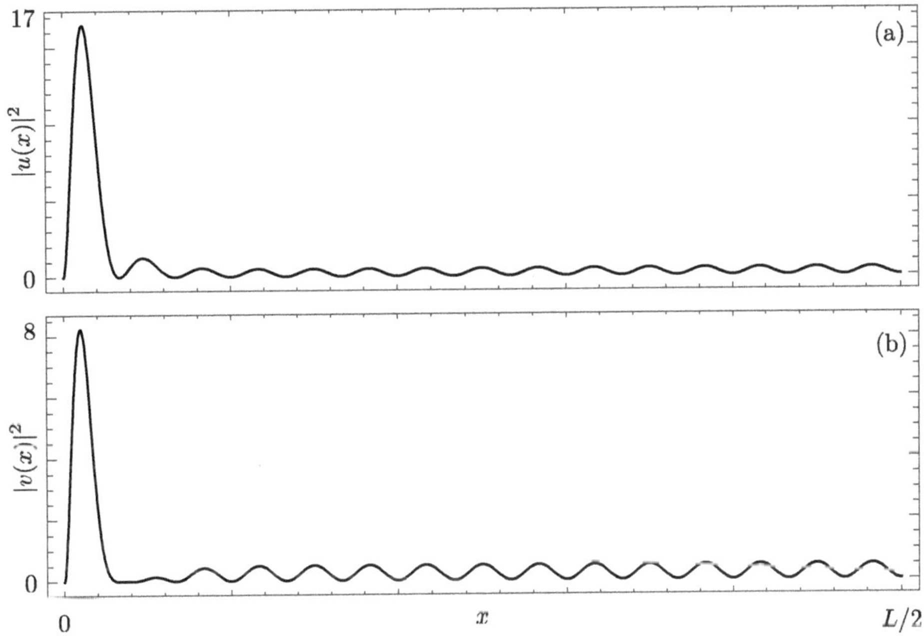

Figure 7. Plot of the probability amplitude of particle and hole components of the zero-mode wavefunction (units L^{-1}) as a function of x: $|u(x)|^2$ (**a**); $|v(x)|^2$ (**b**). Here, $B = \Delta = 30v_F/L$ and $k_F = 30\pi/L$.

A natural question hence arises: are the Majorana zero modes, in the superconducting-dominated phase $B < \Delta$, influenced by the persistence of the Tamm state across the quantum phase transition?

4. Majorana Polarization

The Majorana polarization P_M [64] is a quantity that has been proposed in order to discriminate between Majorana fermions and other types of Andreev bound states. We now evaluate the Majorana polarization associated to the state $\psi_0(x)$, defined as

$$P_M = \frac{N_L}{D_L}, \tag{16}$$

where

$$N_L = \int_0^{L/2} 2u(x)v(x)dx; \quad D_L = \int_0^{L/2} |u(x)|^2 + |v(x)|^2 dx. \tag{17}$$

This absolute value of P_M is close to one (in the large gap limit) if the bound state is a Majorana bound state, while it reaches different values for regular Andreev bound states, which allows us to assess the character of $\psi_0(x)$. In the case under examination, we find

$$P_M = \frac{\left\{L^2k_F^2 + \left[\frac{L(B-\Delta)}{v_F}\right]^2\right\}\left[1+e^{\frac{L(B-\Delta)}{v_F}}\right] - \left\{L^2k_F^2 + \left[\frac{L(B+\Delta)}{v_F}\right]^2\right\}\left[1+e^{-\frac{L(B-\Delta)}{v_F}}\right]}{\left\{L^2k_F^2 + \left[\frac{L(B-\Delta)}{v_F}\right]^2\right\}\left[1+e^{\frac{L(B-\Delta)}{v_F}}\right] + \left\{L^2k_F^2 + \left[\frac{L(B+\Delta)}{v_F}\right]^2\right\}\left[1+e^{-\frac{L(B+\Delta)}{v_F})}\right]}. \quad (18)$$

Looking at the previous expressions, two limiting cases can be extracted. For $B = 0$, one finds $P_M = \tanh\left(\frac{L\Delta}{v_F}\right)$, with $P_M \to 1$ when $\Delta \to \infty$. For $\Delta \ll B$ (and $\mu_0 >> B$), we find $P_M \to -\frac{2B\Delta}{\mu_0^2}$, and thus $P_M \to 0^-$. From the point of view of the Majorana polarization, we hence have that, in the phase dominated by the magnetic field, the Majorana polarization is small, while it approaches, in modulus, the unity in the large gap limit. A transition takes place for $B \simeq \Delta$. However, such a transition does not happen at the exact value of the gap closing, even for $L \to \infty$, that is $B = \Delta$. This behavior of P_M shows, despite the fact that the Majorana phase indeed hosts polarized Majorana fermions, that a reminiscence of the Tamm state is, however, present—and visible—for $B = \Delta$.

The Majorana polarization is shown in Figure 8.

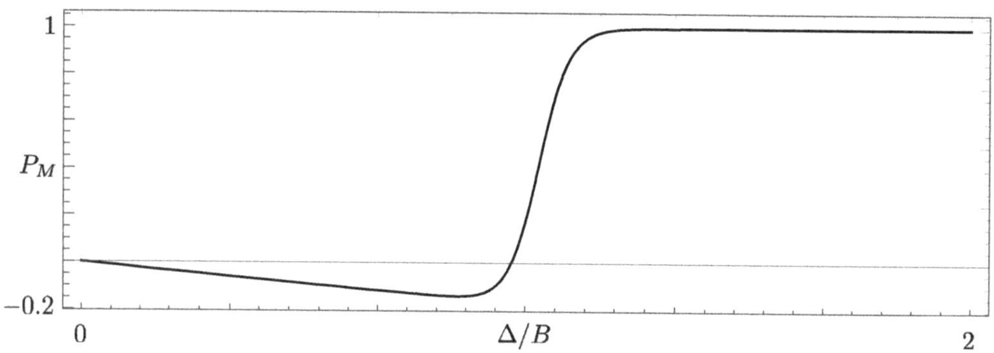

Figure 8. Plot of the Majorana polarization P_M as a function of Δ/B for $k_F = 30\pi/L$ and $B = 30v_F/L$.

5. Conclusions

In this work, we have characterized a simple model for a finite-size, one-dimensional topological superconductor in the presence of a competing normal gapping mechanism arising from a position-dependent potential. The first part of the article deals with the numerical inspection of the model. We find that the model has two phases: one characterized by Majorana fermions at both ends, and the other by, at most, a single Tamm bound state localized at one side. The energy of the bound state, when present, strongly depends on the phase of the position-dependent potential. Surprisingly, we find that the wavefunction of the bound state remains strongly peaked close to the edges, even at the gap closing point separating the two phases. In the second part of the article, we develop an exactly solvable model for interpreting the results of the first part. Finally, in the last section, we evaluate the Majorana polarization for a fully developed bound state of the model and show that, despite the fact that the Tamm state is not completely delocalized at the transition point, deep in the superconducting phase, the Majorana bound states are not strongly affected by the position-dependent potential.

Author Contributions: Conceptualization, M.C., N.T.Z., and F.C.; methodology, L.V.; software, M.C. and F.C.; validation, F.C., M.C., and L.V.; formal analysis, L.V. and N.T.Z.; writing—original draft preparation, N.T.Z. and F.C. All authors have read and agreed to the published version of the manuscript.

Funding: This research received no external funding.

Institutional Review Board Statement: Not applicable.

Informed Consent Statement: Not applicable.

Data Availability Statement: Not applicable.

Acknowledgments: We acknowledge fruitful discussions with Simone Traverso and Maura Sassetti.

Conflicts of Interest: The authors declare no conflict of interest.

References

1. Nayak, C.; Simon, S.H.; Stern, A.; Freedman, M.; Das Sarma, S. Non-Abelian Anyons and Topological Quantum Computation. *Rev. Mod. Phys.* **2008**, *80*, 1083. [CrossRef]
2. Read, N.; Rezayi, E. Beyond paired quantum hall states: Parafermions and incompressible states in the first excited landau level. *Phys. Rev. B* **1999**, *59*, 8084. [CrossRef]
3. Blasi, A.; Braggio, A.; Carrega, M.; Ferraro, D.; Maggiore, N.; Magnoli, N. Non-Abelian BF theory for 2+1 dimensional topological states of matter. *New J. Phys.* **2012**, *14*, 013060. [CrossRef]
4. Zhang, F.; Kane, C.L.Time-Reversal-Invariant Z_4 Fractional Josephson Effect *Phys. Rev. Lett.* **2014**, *113*, 036401. [CrossRef] [PubMed]
5. Orth, C.P.; Tiwari, R.P.; Meng, T.; Schmidt, T.L. Non-Abelian parafermions in time-reversal-invariant interacting helical systems. *Phys. Rev. B* **2015**, *91*, 081406. [CrossRef]
6. Klinovaja, J.; Loss, D. Fractional charge and spin states in topological insulator constrictions. *Phys. Rev. B* **2015**, *92*, 121410. [CrossRef]
7. Fleckenstein, C.; Traverso Ziani, N.; Trauzettel, B. Z_4 parafermions in Weakly Interacting Superconducting Constrictions at the Helical Edge of Quantum Spin Hall Insulators. *Phys. Rev. Lett.* **2020**, *122*, 066801. [CrossRef] [PubMed]
8. Rossini, D.; Carrega, M.; Calvanese Strinati, M.; Mazza, L. Anyonic tight-binding models of parafermions and of fractionalized fermions. *Phys. Rev. B* **2019**, *99*, 085113. [CrossRef]
9. Carrega, M.; Principi, A.; Vera-Marun, I. Tunneling spectroscopy as a probe of fractionalization in 2D magnetic heterostructures. *Phys. Rev. B* **2020**, *102*, 085412. [CrossRef]
10. Ronetti, F.; Carrega, M.; Ferraro, D.; Rech, J.; Jonckheere, T.; Martin, T.; Sassetti, M. Polarized heat current generated by quantum pumping in two-dimensional topological insulators. *Phys. Rev. B* **2017**, *95*, 115412. [CrossRef]
11. Schiller, N.; Cornfeld, E.; Berg, E.; Oreg, Y. Predicted signatures of topological superconductivity and parafermion zero modes in fractional quantum Hall edges. *Phys. Rev. Res.* **2020**, *2*, 023296. [CrossRef]
12. Michelsen, A.B.; Schmidt, T.L.; Idrisov, E.G. Current correlations of Cooper-pair tunneling into a quantum Hall system. *Phys. Rev. B* **2020**, *102*, 125402. [CrossRef]
13. Zhao, L.; Arnault, E.G.; Bondarev, A.; Seredinski, A.; Larson, T.F.Q.; Draelos, A.W.; Li, H.; Watanabe, K.; Taniguchi, T.; Amet, F.; et al. Interference of chiral Andreev edge states, *Nat. Phys.* **2020**, *16*, 862. [CrossRef]
14. Amet, F.; Ke, C.T.; Borzenets, I.V.; Wang, J.; Watanabe, K.; Taniguchi, T.; Deacon, R.S.; Yamamoto, M.; Bomze, Y.; Tarucha, S.; et al. Supercurrent in the quantum Hall regime. *Science* **2016**, *352*, 966. [CrossRef] [PubMed]
15. Lee, G.-H.; Huang, K.-F.; Efetov, D.K.; Wei, D.S.; Hart, S.; Taniguchi, T.; Watanabe, K.; Yacoby, A.; Kim, P. Inducing Superconducting Correlation in Quantum Hall Edge States. *Nat. Phys.* **2017**, *13*, 693. [CrossRef]
16. Gül, O.; Ronen, Y.; Lee, S.Y.; Shapourian, H.; Zauberman, J.; Lee, Y. H.; Watanabe, K.; Taniguchi, T.; Vishwanath, A.; Yacoby, A.; et al. Induced superconductivity in the fractional quantum Hall edge. *arXiv*, **2020**, arXiv:2009.07836.
17. Mourik, V.; Zuo, K.; Frolov, S.M.; Plissard, S.R.; Bakkers, E.P.A.M.; Kouwenhoven, L.P. Signatures of Majorana fermions in hybrid superconductor-semiconductor nanowire devices. *Science* **2012**, *336*, 1003. [CrossRef]
18. Lutchyn, R. M.; Sau, J. D.; Das Sarma, S. Majorana Fermions and a Topological Phase Transition in Semiconductor-Superconductor Heterostructures. *Phys. Rev. Lett.* **2010**, *105*, 077001. [CrossRef]
19. Oreg, Y.; Refael, G.; von Oppen, F. Helical Liquids and Majorana Bound States in Quantum Wires. *Phys. Rev. Lett.* **2010**, *105*, 177002. [CrossRef] [PubMed]
20. Fu, L.; Kane, C.L. Superconducting proximity effect and Majorana fermions at the surface of a topological insulator. *Phys. Rev. Lett.* **2008**, *100*, 096407. [CrossRef] [PubMed]
21. Fu, L.; Kane, C.L. Josephson current and noise at a superconductor/quantum-spin-Hall-insulator/superconductor junction. *Phys. Rev. B* **2009**, *79*, 161408. [CrossRef]
22. Akhmerov, A.R.; Nilsson, J.; Beenakker, C.W.J. Electrically detected interferometry of Majorana fermions in a topological insulator. *Phys. Rev. Lett.* **2009**, *102*, 216404. [CrossRef]

23. Strunz, J.; Wiedenmann, J.; Fleckenstein, C.; Lunczer, L.; Beugeling, W.; Müller, V.L.; Shekhar, P.; Traverso Ziani, N.; Shamim, S.; Kleinlein, J.; et al. Interacting topological edge channels. *Nat. Phys.* **2020**, *16*, 83. [CrossRef]
24. Li, J.; Pan, W.; Bernevig, B.A.; Lutchyn, R.M. Detection of Majorana Kramers Pairs Using a Quantum Point Contact. *Phys. Rev. Lett.* **2016**, 117, 046804. [CrossRef] [PubMed]
25. Fleckenstein, C.; Traverso Ziani, N.; Calzona, A.; Sassetti, M.; Trauzettel, B. Formation and detection of Majorana modes in quantum spin Hall trenches. *Phys. Rev. B* **2021**, *103*, 125303. [CrossRef]
26. Nadj-Perge, S.; Drozdov, I.K.; Li, J.; Chen, H.; Jeon, S.; Seo, J.; MacDonald, A.H.; Bernevig, B.A.; Yazdani, A. Observation of Majorana fermions in ferromagnetic atomic chains on a superconductor. *Science* **2014**, *6209*, 602. [CrossRef]
27. Ren, H.; Pientka, F.; Hart, S.; Pierce, A.T.; Kosowsky, M.; Lunczer, L.; Schlereth, R.; Scharf, B.; Hankiewicz, E.M.; Molenkamp, L.W.; et al. Topological superconductivity in a phase-controlled Josephson junction. *Nature* **2019**, *569*, 93. [CrossRef] [PubMed]
28. Guiducci, S.; Carrega, M.; Biasiol, G.; Sorba, L.; Beltram, F.; Heun, S. Toward Quantum Hall Effect in a Josephson Junction. *Phys. Status Solidi RRL* **2019**, *13*, 1800222. [CrossRef]
29. Guiducci, S.; Carrega, M.; Taddei, F.; Biasiol, G.; Courtois, H.; Beltram, F.; Heun, S. Full electrostatic control of quantum interference in an extended trenched Josephson junction. *Phys. Rev. B* **2019**, *99*, 235419. [CrossRef]
30. Krogstrup, P.; Ziino, N.L.B.; Chang, W.; Albrecht, S.M.; Madsen, M.H.; Johnson, E.; Nygard, J.; Marcus, C.M.; Jespersen, T.S. Epitaxy of semiconductor–superconductor nanowires. *Nat. Mater.* **2015**, *14*, 400. [CrossRef]
31. Deng, M.T.; Vaitiekenas, S.; Hansen, E.B.; Danon, J.; Leijnse, M.; Flensberg, K.; Nygard, J.; Krogstrup, P.; Marcus, C.M. Majorana bound state in a coupled quantum-dot hybrid-nanowire system. *Science* **2016**, *354*, 1557. [CrossRef]
32. Aguado, R. Majorana quasiparticles in condensed matter. *La Riv. Del Nuovo Cimento* **2017**, *40*, 523.
33. Asano, Y.; Tanaka, Y.; Kashiwaya, S. Phenomenological theory of zero-energy Andreev resonant states. *Phys. Rev. B* **2004**, *69*, 134501. [CrossRef]
34. Tanaka, Y.; Kashiwaya, S.; Yokoyama, T. Theory of enhanced proximity effect by midgap Andreev resonant state in diffusive normal-metal/triplet superconductor junctions. *Phys. Rev. B* **2005**, *71*, 094513. [CrossRef]
35. Golubov, A.A.; Brinkman, A.; Tanaka, Y.; Mazin, I.I.; Dolgov, O.V. Andreev Spectra and Subgap Bound States in Multiband Superconductors. *Phys. Rev. Lett.* **2009** *103*, 077003. [CrossRef]
36. Tanaka, Y.; Mizuno, Y.; Yokoyama, T.; Yada, K.; Sato, M. Anomalous Andreev Bound State in Noncentrosymmetric Superconductors. *Phys. Rev. Lett.* **2012**, *105*, 097002. [CrossRef] [PubMed]
37. Liu, J.; Potter, A.C.; Law, K.T.; Lee, P.A. Zero-Bias Peaks in the Tunneling Conductance of Spin-Orbit-Coupled Superconducting Wires with and without Majorana End-States. *Phys. Rev. Lett.* **2012**, *109*, 267002. [CrossRef] [PubMed]
38. Kells, G.; Meidan, D.; Brouwer, P.W. Near-zero-energy end states in topologically trivial spin-orbit coupled superconducting nanowires with a smooth confinement. *Phys. Rev. B* **2012**, *86*, 100503. [CrossRef]
39. Roy, D.; Bondyopadhaya, N.; Tewari, S. Topologically trivial zero-bias conductance peak in semiconductor Majorana wires from boundary effects, *Phys. Rev. B* **2013**, *88*, 020502. [CrossRef]
40. Stanescu, T.D.; Tewari, S. Disentangling Majorana fermions from topologically trivial low-energy states in semiconductor Majorana wires. *Phys. Rev. B* **2013** *87*, 140504. [CrossRef]
41. Cayao, J.; Prada, E.; San-Jose, P.; RAguado, R. SNS junctions in nanowires with spin-orbit coupling: Role of confinement and helicity on the subgap spectrum. *Phys. Rev. B* **2015**, *91*, 024514. [CrossRef]
42. San-Jose, P.; Cayao, J.; Prada, E.; Aguado, R. Majorana bound states from exceptional points in non-topological superconductors. *Sci. Rep.* **2016**, *6*, 21427. [CrossRef]
43. Liu, C.-X.; Sau, J.D.; Stanescu, T.D.; Das Sarma, S. Andreev bound states versus Majorana bound states in quantum dot-nanowire-superconductor hybrid structures: Trivial versus topological zero-bias conductance peaks. *Phys. Rev. B* **2017**, *96*, 075161. [CrossRef]
44. Liu, C.-X.; Sau, J.D.; Das Sarma, S. Distinguishing topological Majorana bound states from trivial Andreev bound states: Proposed tests through differential tunneling conductance spectroscopy. *Phys. Rev. B* **2018**, *97*, 214502. [CrossRef]
45. Moore, C.; Stanescu, S.T.; Tewari, S. Two-terminal charge tunneling: Disentangling Majorana zero modes from partially separated Andreev bound states in semiconductorsuperconductor heterostructures. *Phys. Rev. B* **2018**, *97*, 165302. [CrossRef]
46. Moore, C.; Zeng, C.; Stanescu, ; T.D.; Tewari, S. Quantized zero-bias conductance plateau in semiconductorsuperconductor heterostructures without topological Majorana zero modes. *Phys. Rev. B* **2018**, *98*, 155314. [CrossRef]
47. Fleckenstein, C.; Dominguez, F.; Traverso Ziani, N.; Trauzettel, B. Decaying spectral oscillations in a Majorana wire with finite coherence length. *Phys. Rev. B* **2018**, *97*, 155425. [CrossRef]
48. Awoga, O.A.; Cayao, J.; Black-Schaffer, A.M. Supercurrent Detection of Topologically Trivial Zero-Energy States in Nanowire Junctions. *Phys. Rev. Lett.* **2019**, *123*, 117001. [CrossRef] [PubMed]
49. Marra, P.; Nitta, M. Topologically nontrivial Andreev bound states. *Phys. Rev. B* **2019**, *100*, 220502. [CrossRef]
50. Cayao, J.; Black-Schaffer, A.M. Distinguishing trivial and topological zero energy states in long nanowire junctions. *arXiv* **2020**, arXiv:2011.10411.
51. Pan, H.; Das Sarma, S. Physical mechanisms for zero-bias conductance peaks in Majorana nanowires. *Phys. Rev. Res.* **2020**, *2*, 013377. [CrossRef]

52. Huang, Y.; Pan, H.; Liu, C.-X.; Sau, J.D.;Stanescu, T.D.; Das Sarma, S. Metamorphosis of Andreev bound states into Majorana bound states in pristine nanowires. *Phys. Rev. Lett.* **2018**, *98*, 144511. [CrossRef]
53. Gangadharaiah, S.; Trifunovic, L.; Loss, D. Localized End States in Density Modulated Quantum Wires and Rings. *Phys. Rev. Lett.* **2012**, *108*, 136803. [CrossRef] [PubMed]
54. Traverso Ziani, N.; Fleckenstein, C.; Vigliotti, L.; Trauzettel, B.; Sassetti, M. From fractional solitons to Majorana fermions in a paradigmatic model of topological superconductivity. *Phys. Rev. B* **2020**, *101*, 195303. [CrossRef]
55. Henriques, J.C.G.; Rappoport, T.G.; Bludov, Y.V.; Vasilevskiy, M.I.; Peres, N.M.R. Topological photonic Tamm states and the Su-Schrieffer-Heeger model. *Phys. Rev. A* **2020**, *101*, 043811. [CrossRef]
56. Jackiw, R.; Rebbi, C. Solitons with fermion number ½. *Phys. Rev. D* **1976**, *13*, 3398. [CrossRef]
57. Goldstone, J.; Wilczek, F. Fractional Quantum Numbers on Solitons. *Phys. Rev. Lett.* **1981**, *47*, 986. [CrossRef]
58. Kivelson, S.; Schrieffer, J.R. Wannier functions in one-dimensional disordered systems: Application to fractionally charged solitons. *Phys. Rev. B* **1982**, *25*, 6447. [CrossRef]
59. Qi, X.-L.; Hughes, T. L.; Zhang, S.-C. Fractional charge and quantized current in the quantum spin Hall state. *Nat. Phys.* **2008**, *4*, 273. [CrossRef]
60. Väyrynen J.I.; Ojanen, T. Chiral Topological Phases and Fractional Domain Wall Excitations in One-Dimensional Chains and Wires. *Phys. Rev. Lett.* **2011**, *107*, 166804. [CrossRef] [PubMed]
61. Klinovaja, J.; Stano, P.; Loss, D. Transition from Fractional to Majorana Fermions in Rashba Nanowires. *Phys. Rev. Lett.* **2012**, *109*, 236801. [CrossRef]
62. Fleckenstein, C.; Traverso Ziani, N.; Trauzettel, B. Chiral anomaly in real space from stable fractional charges at the edge of a quantum spin Hall insulator. *Phys. Rev. B* **2016**, *94*, 241406. [CrossRef]
63. Gresta, D.; Blasi, G.; Taddei, F.; Carrega, M.; Braggio, A.; Arrachea, L. Signatures of Jackiw-Rebbi resonance in the thermal conductance of topological Josephson junctions with magnetic islands. *Phys. Rev. B* **2021**, *103*, 075439. [CrossRef]
64. Sticlet, D.; Bena, C.; Simon, P. Spin and Majorana Polarization in Topological Superconducting Wires. *Phys. Rev. Lett.* **2012**, *108*, 096802. [CrossRef] [PubMed]
65. Sedlmayr, N.; Bena, C. Visualising Majorana bound states in 1D and 2D using the generalized Majorana polarization. *Phys. Rev. B* **2015**, *92*, 115115. [CrossRef]
66. Bena, C. Testing the formation of Majorana states using Majorana polarization. *Comptes Rendus Phys.* **2017**, *18*, 349. [CrossRef]
67. Bernevig, B.A.; Hughes, T.L. *Topological Insulators and Topological Superconductors*; Princeton University Press: Princeton, NJ, USA, 2013.
68. Malard, M.; Japaridze, G. I.; Johannesson, H. Synthesizing Majorana zero-energy modes in a periodically gated quantum wire. *Phys. Rev. B* **2016**, *94*, 115128. [CrossRef]
69. Peierls, R.A. Zur Theorie der elektrischen und thermischen Leitfähigkeit von Metallen. *Ann. Phys.* **1930**, *4*, 121. [CrossRef]
70. Haldane, F.D.M. Effective harmonic-fluid approach to low-energy properties of one-dimensional quantum fluids. *Phys. Rev. Lett.* **1981**, *47*, 1840. [CrossRef]
71. Cuniberti, G.; Sassetti, M.; Kramer, B. Transport and elementary excitations of a Luttinger liquid. *J. Phys. Condens. Matter* **1996**, *8*, L21. [CrossRef]
72. Guinea, F.; Santos, G.G.; Sassetti, M.; Ueda, M. Asymptotic tunneling conductance in Luttinger liquids. *Europhys. Lett.* **80**, *30*, 561. [CrossRef]
73. Fabrizio, M.; Gogolin, A.O. Interacting one-dimensional electron gas with open boundaries. *Phys. Rev. B* **1995**, *51*, 17827. [CrossRef] [PubMed]
74. Traverso Ziani, N.; Cavaliere, F.; Sassetti, M. Signatures of Wigner correlations in the conductance of a one-dimensional quantum dot coupled to an AFM tip. *Phys. Rev. B* **2012**, *86*, 125451. [CrossRef]
75. Traverso Ziani, N.; Cavaliere, F.; Sassetti, M. Theory of the STM detection of Wigner molecules in spin-incoherent CNTs. *Europhys. Lett.* **2013**, *102*, 47006. [CrossRef]
76. Porta, S.; Gambetta, F.M.; Traverso Ziani, N.; Kennes, D.M.; Sassetti, M.; Cavaliere, F. Nonmonotonic response and light-cone freezing in fermionic systems under quantum quenches from gapless to gapped or partially gapped states. *Phys. Rev. B* **2018**, *97*, 035433. [CrossRef]
77. Giamarchi T. *Quantum Physics in One Dimension;* Oxford University Press: Oxford, UK, 2004; ISBN 9780198525004.
78. Timm, C. Transport through a quantum spin Hall quantum dot. *Phys. Rev. B* **2012**, *86*, 155456. [CrossRef]
79. Dolcetto, G.; Traverso Ziani, N.; Biggio, M.; Cavaliere, F.; Sassetti, M. Spin textures of strongly correlated spin Hall quantum dots. *Phys. Stat. Sol. (RRL)* **2013**, *7*, 1059. [CrossRef]
80. Kylänpää, I.; Cavaliere, F.; Traverso Ziani, N.; Sassetti, M.; Räsänen, E. Thermal effects on the Wigner localization and Friedel oscillations in many-electron nanowires. *Phys. Rev. B* **2016**, *94*, 115417. [CrossRef]
81. Alicea, J. New directions in the pursuit of Majorana fermions in solid state systems. *J. Rep. Prog.* **2012**, *75*, 076501. [CrossRef] [PubMed]
82. Franchini, F. *An Introduction to Integrable Techniques for One-Dimensional Quantum Systems;* Lecture Notes in Physics; Springer: Berlin, Germany, 2017; Volume 940.
83. Porta, S.; Cavaliere, F.; Sassetti, M.; Ziani, N.T. Topological classification of dynamical quantum phase transitions in the xy chain. *Sci. Rep.* **2020**, *10*, 642. [CrossRef]

symmetry

MDPI

Review

Topological Excitations in Neutral–Ionic Transition Systems

Keishi Sunami [1,*,†], Ryosuke Takehara [1,‡], Kazuya Miyagawa [1], Hiroshi Okamoto [2] and Kazushi Kanoda [1,*]

1 Department of Applied Physics, University of Tokyo, Bunkyo-ku, Tokyo 113-8656, Japan; takehara.r.ab@m.titech.ac.jp (R.T.); kazuya@ap.t.u-tokyo.ac.jp (K.M.)
2 Department of Advanced Materials Science, University of Tokyo, Kashiwa, Chiba 277-8561, Japan; okamotoh@k.u-tokyo.ac.jp
* Correspondence: k.sunami@aist.go.jp (K.S.); kanoda@ap.t.u-tokyo.ac.jp (K.K.)
† Current address: Research Institute for Advanced Electronics and Photonics, National Institute of Advanced Industrial Science and Technology (AIST), Tsukuba 305-8565, Ibaraki, Japan.
‡ Current address: Laboratory for Chemistry and Life Science, Institute of Innovative Research, Tokyo Institute of Technology, Yokohama 226-8503, Kanagawa, Japan.

Abstract: The existence and physical properties of topological excitations in ferroelectrics, especially mobile topological boundaries in one dimension, are of profound interest. Notably, topological excitations emerging in association with the neutral–ionic (NI) phase transition are theoretically suggested to carry fractional charges and cause anomalous charge transport. In recent years, we experimentally demonstrated mobile topological excitations in a quasi-one-dimensional (1D) ferroelectric, tetrathiafulvalene-*p*-chloranil [TTF-CA; TTF ($C_6H_4S_4$) and CA ($C_6Cl_4O_2$)], which shows the NI transition, using NMR, NQR, and electrical resistivity measurements. Thermally activated topological excitations carry charges and spins in the NI crossover region and in the ionic phase with a dimer liquid. Moreover, free solitons show a binding transition upon a space-inversion symmetry-breaking ferroelectric order. In this article, we review the recent progress in the study of mobile topological excitations emerging in TTF-CA, along with earlier reports that intensively studied these phenomena, aiming to provide the foundations of the physics of electrical conductivity and magnetism carried by topological excitations in the 1D ferroelectric.

Keywords: topological excitation; domain wall; soliton; neutral–ionic transition; electronic ferroelectric

1. Introduction

The concept of topology has opened a new window in condensed matter physics for identifying novel electronic states and phenomena [1–6]. Among them, boundaries dividing topologically different phases, such as surfaces for three-dimensional (3D) systems and edges for two-dimensional systems, are intensively researched fields, where peculiar electronic states emerge, such as gapless surface bands [6]. In the studies to date, however, the topological boundaries are built-in systems and remain static. If the topological boundaries are mobile, they are expected to offer new degrees of freedom and open a new phase in the development of topological condensed matter physics. In one-dimensional (1D) systems, the boundaries are in the form of zero-dimensional (0D) points, which can be regarded as a kind of particle and therefore have the potential to be mobile, unlike surfaces and lines in higher dimensions. The representative example of topological excitation in 1D systems is a soliton emerging in a conducting polymer, polyacetylene, in which the charged solitons are responsible for high electrical conductivity [3].

Interestingly, the 1D electronic ferroelectric, in which charge, spin, and lattice degrees of freedom are strongly entangled, is suggested to host the solitons as elementary excitations carrying a fractional charge and unusual conducting properties. The neutral–ionic (NI) transition material tetrathiafulvalene-*p*-chloranil (TTF-CA), which is a quasi-1D ferroelectric, has long been argued to be a promising candidate for the emergence of such topological excitations [7–15]; however, the existence and physical properties of thermally excited solitons as suggested theoretically remain elusive. Recently, experimental studies on topological excitations in the NI transition system have made remarkable progress because the phase diagram, revisited with experiments performed in a regime with higher pressure than ever, has proved to have features that can host topological excitations at high temperatures [16–19]; more specifically, the mobile topological spin and charge excitations, namely, spin solitons and the neutral–ionic domain wall (NIDW), are captured by nuclear magnetic resonance (NMR) and electrical resistivity measurements. Such excitations lead to a novel mechanism of electrical conduction. Moreover, the solitons show a binding transition upon the long-range ferroelectric order, which is a unique phenomenon that has never been observed in conducting polymers.

In this review article, we report the recent progress in the physics of mobile topological excitations in TTF-CA. In Section 2, we briefly introduce the fundamental properties of the NI transition material TTF-CA and the concept of topological excitations emerging in this system. In Section 3, we describe the pressure–temperature (*P*-*T*) phase diagram of TTF-CA revisited by nuclear quadrupole resonance (NQR) and highlight high-temperature characteristics revised from an earlier report. In Section 4, we review the experimental studies on electrical conductivity and magnetism brought about by mobile topological excitations in TTF-CA; the paramagnetism carried by mobile spin solitons (Section 4.1); the anomalous charge transport caused by mobile NIDWs and solitons (Section 4.2); and the binding transition of solitons upon the space-inversion symmetry-breaking ferroelectric order (Section 4.3). In Section 5, we summarize this review.

2. Topological Excitations in Neutral–Ionic Transition Systems

2.1. Neutral–Ionic Transition Material TTF-CA

The quasi 1D donor–acceptor (DA) complex, TTF-CA, is a strongly charge–spin–lattice coupled system, which is the representative NI transition material [20–30]. Molecular arrangements in TTF-CA are shown in Figure 1a,b. The TTF and CA molecules alternatingly stack face-to-face along the *a*-axis, forming 1D chains. At ambient temperature and pressure, TTF-CA is in a quasi-neutral (N) state; the highest occupied molecular orbitals (HOMOs) of the donor molecule, TTF, and the acceptor molecule, CA, are doubly occupied by electrons. Upon cooling, the system transitions to a quasi-ionic (I) state at 81 K with an electron transfer from the HOMO of TTF to the lowest unoccupied molecular orbital (LUMO) of CA due to the Madelung energy gain (see Figure 2) [21,23]. The Madelung energy incorporates long-range Coulomb interactions between all of the molecules in the crystal, and thus, the charge transfer upon the NI transition is a collective phenomenon. The N and I phases are characterized by the degree of charge transfer ρ defined by $D^{+\rho}A^{-\rho}$: $\rho_N \sim 0.3$ in the N phase and $\rho_I \sim 0.6$–0.7 in the I phase [24,31–35]. The partial charge transfer from TTF to CA is due to the hybridization of the molecular orbitals between them.

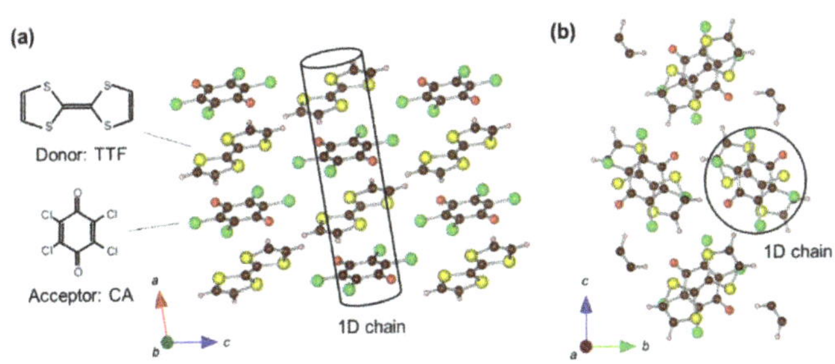

Figure 1. Molecular arrangements in the *ac* plane (**a**) and *bc* plane (**b**) in the neutral phase of TTF-CA.

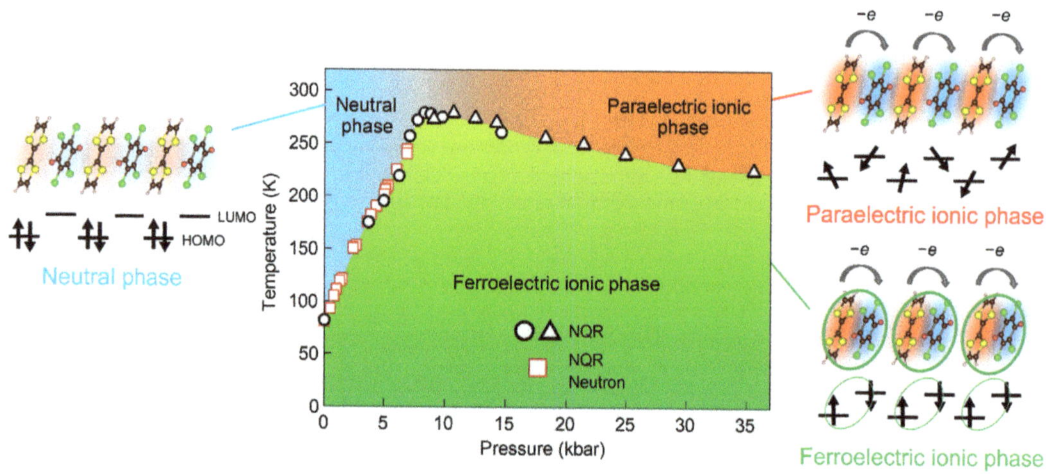

Figure 2. Pressure–temperature phase diagram of TTF-CA. Black open circles (sample #1) and triangles (sample #2) represent the phase transition temperatures T_c determined by the NQR experiments in [16]. Red open squares represent T_c determined by the NQR and neutron experiments in [36]. Reprinted with permission from [16]. Copyright 2018, American Physical Society. Schematics in the sides of the central panel depict the electron configurations in space and energy in TTF and CA molecules; red and blue clouds represent the hole and electron densities in HOMO of TTF and LUMO of CA by color intensity. Reprinted with permission from [18]. Copyright 2019, The Authors, some rights reserved; exclusive licensee AAAS. Distributed under a CC BY-NC 4.0 License (http://creativecommons.org/licenses/by-nc/4.0/ (accessed on 15 November 2021)).

Lattice dimerization is caused in the charge-transferred I phase due to the Peierls or spin-Peierls instability inherent in 1D systems [25,37–39]. Thus, the 1D dimerized chains are polarized, which show a long-range ferroelectric order in the I phase. The ferroelectricity in TTF-CA is of keen interest because the charge transfer coupled with the dimerization yields electronic ferroelectricity [40–43], which can be described by a modern ferroelectric theory in the framework of a Berry phase [44–47].

The charge transfer upon the NI transition generates an $S = 1/2$ spin on each molecule in the I phase, but the spins form singlet pairs due to the dimerization. Thus, TTF-CA is a nonmagnetic insulator in both the N phase and ionic ferroelectric (I_{ferro}) phase at ambient pressure: a band insulator with a closed-shell structure in the former phase and an ionic Mott insulator with a spin-singlet formation in the latter phase. In the following sections, we review studies showing that the entanglement of charge, spin, and lattice degrees of freedom brings about emergent excitations with topological character in TTF-CA.

2.2. Types of Topological Excitations in Neutral–Ionic Transition Systems

The concept of topological excitations in the NI transition systems was theoretically proposed in the 1980s [7,48,49] as defects that emerge in between domains of degenerate ground states and thus are activated near the region where more than two phases compete. In the NI transition system of TTF-CA, the N and I_{ferro} phases are degenerate near the NI phase boundary due to the coupling between charge-transfer and lattice instabilities, as mentioned in Section 2.1. Moreover, two opposite polar patterns (denoted by I_A and I_B in Figure 3) are degenerate in the I_{ferro} phase. Thus, it is theoretically proposed that neutral–ionic domain walls (NIDWs) reside in between the N and I_A (or I_B) domains, and spin and spinless solitons are located in between the I_A and I_B domains (see Figure 3). The spinless soliton is often called a charge soliton, which corresponds to a single neutral molecule sandwiched by two NIDWs. The difference between spin and charge solitons comes from the number of electrons that occupy the molecular orbital of the soliton site; for the spin soliton, the HOMO of D or the LUMO of A is occupied by one electron, whereas, for the charge soliton, the HOMO of D is fully occupied or the LUMO of A is empty. The topological nature of these excitations is expressed based on the phase Hamiltonian in [7,10]. Physically, phases of charge, spin, and lattice modulations have discontinuities at the NI domain walls, spin solitons, and charge solitons; these discontinuities that are characterized by topological charge or spin are robust and cannot annihilate without pair collision. These excitations have fractional charges: $e(\rho_I - \rho_N)/2$ for the NIDW, $e(1 - \rho_I)$ for the spin soliton, and $-e\rho_I$ for the charge soliton, where e is the elementary charge. Remarkably, the NIDW is freely mobile near the NI boundary where the N and I states are degenerate, and its excitation energy is lowered due to the itinerant nature, attributed to the large transfer integral along the 1D direction [48]. Thus, NIDWs can be low-energy excitations in NI transition systems uniquely and are expected to yield unconventional charge transport that differs from that of individual electrons, as theoretically suggested [7,9,10]; note that both the NIDW (or charge soliton) and spin soliton are necessary for carrying a steady electrical current, the mechanism of which is briefly explained in Section 4.2.

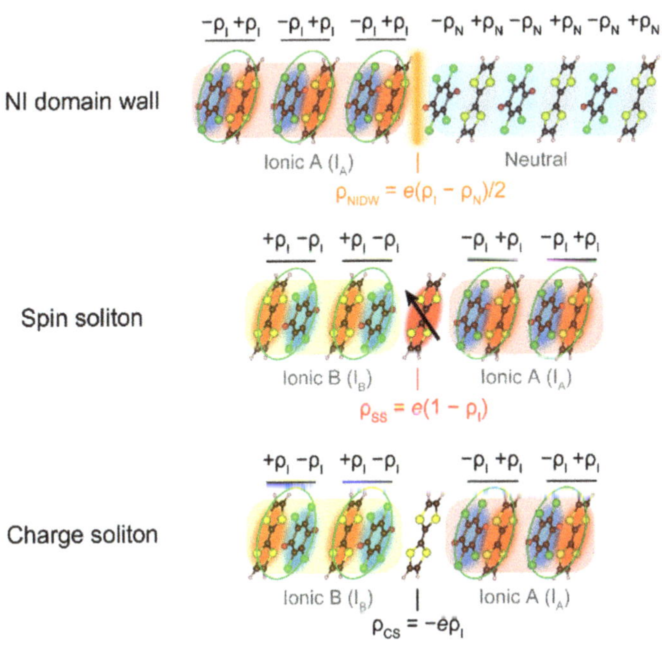

Figure 3. Topological excitations in neutral–ionic phase transition systems. ρ_{NIDW}, ρ_{SS}, and ρ_{CS} are topological charges of a NI domain wall, a spin soliton, and a charge soliton, respectively. ρ_N and ρ_I represent the degree of charge transfer from HOMO of TTF and LUMO of CA in the neutral and ionic phases, respectively. Modified with permission from [18]. Copyright 2019, The Authors, some rights reserved; exclusive licensee AAAS. Distributed under a CC BY-NC 4.0 License (http://creativecommons.org/licenses/by-nc/4.0/ (accessed on 15 November 2021)).

In the past, these topological boundaries were observed as frozen defects in the ferroelectric phase and its vicinity. The spin solitons were detected in the I_{ferro} phase at ambient pressure by ESR measurements [11,50], which were generated in the multi-domain structure of the I_{ferro} phase (the typical ferroelectric domain size was reported to be ~100 μm [51]) when cooled below the transition temperature T_c. Thus, they behave as frozen defects, not thermally activated, in the 3D ferroelectric order. The density of the thus-generated spin soliton was estimated to be ~10^{-4} spins per molecule from the absolute value of spin susceptibility, which is extremely dilute. The emergence of charge solitons was suggested by dielectric measurements [15,50], which found an anomalous enhancement of the dielectric constant in the I_{ferro} phases with a characteristic frequency dependence in a frequency range below ~MHz. It was attributed to the motions of charge solitons. Enhanced dielectric constants in an analogous material, TTF-$QBrCl_3$, are suppressed by electric-field poling, which varies the density of frozen charge solitons in the ferroelectric phase [50], indicating the involvement of domain boundaries in enhanced dielectric constants.

In the N phase near the NI transition temperature, fluctuating I domains sandwiched by two NIDWs with charges of opposite signs were detected [52,53]. Their motion is suggested to contribute to the enhancement of dielectric constants in the N phase [15], as with the charge solitons in the I_{ferro} phase. Moreover, the NIDW motion also pertains to the photoinduced phase transition (PIPT) of TTF-CA, which has long attracted much attention as a prototypical system of PIPT [29,54–58]. The N and I phases can be switched by photoirradiation; recent studies have shown that the electric polarization is controllable using the THz electric-field pulse, causing the displacement of the NIDW [53,59].

Regarding the charge transport, intriguing phenomena have been observed [12,14]; the two-terminal resistivity decreases around the NI transition under ambient pressure in TTF-CA despite the fact that the system is an insulator in both the N and I_{ferro} phases, which evokes some unconventional conduction mechanism. However, this decrease is not due to the NIDW or soliton conduction, as was demonstrated by resistivity measurements under the electric-field control of the soliton density in the TTF-$QBrCl_3$ mentioned above [50]; the electrical conductivity increases when the soliton density is suppressed, possibly suggesting that the soliton is the carrier scatterer, rather than the charge carrier.

Why has the topological charge transport been experimentally elusive in the NI transition system? This is very probably because the 3D long-range ferroelectric order prohibits the travel of NIDWs and solitons at ambient pressure (although the displacement current flows due to their transient motions). Thus, if NIDWs are activated in the nonordered paraelectric regime, they may travel and carry a charge current. As reported in Sections 3 and 4, we demonstrated anomalous charge transport by topological excitations under pressures where the N phase and the I phase free from the dimer order, which we call the paraelectric ionic (I_{para}) phase, are temporally fluctuating. Moreover, in the 3D ferroelectric ordered (I_{ferro}) phase, instead of frozen solitons, we revealed that bound pairs of solitons are responsible for charge and spin excitations.

3. Revisited Pressure–Temperature Phase Diagram of TTF-CA

3.1. Previous Phase Diagram

In this section, we describe the physical properties of TTF-CA under pressure. First, we summarize under-pressure studies conducted earlier in TTF-CA using neutron scattering, X-ray diffraction, NQR, optical spectroscopy (using infrared and visible light), dielectric, and electrical conductivity measurements [22,36,60–62]. The NI transition occurs simultaneously with the dimerization transition under pressures of up to ~7 kbar, as detected by NQR and neutron scattering measurements (Figure 2); with increasing pressure, the phase transition temperature, T_c, shows a continuous increase, and the jump in the degree of charge transfer ρ is diminished. Upon a pressure sweep at room temperature, DA dimerization is detected at high pressures by infrared spectroscopy measurements, which show that the intensity of the a_g mode of the CA molecule, which is sensitive to DA dimerization, is rapidly enhanced above 8 kbar and reaches a maximum at 11 kbar [27,63,64]. X-ray scattering showed an anomalous lattice-parameter change in the a-axis direction at pressures of 6.5–11 kbar at 300 K that deviated from the pressure evolution in the lattice parameter in the N and I phases [62]; thus, a ferroelectric transition at 11 kbar was inferred. The electrical conductivity measured by the two-terminal method exponentially increases with pressure and exhibits a peak at ~9 kbar [12,14], which is almost located on the extrapolation of the NI transition line determined by NQR and neutron scattering at lower pressures below

7 kbar. Thus, the inclined NI transition line between the N and I_{ferro} phases is likely to be in the P-T phase diagram.

On the other hand, the temperature profiles of NQR and neutron diffraction data near 7 kbar suggest that the charge transfer may be separated in temperature from that of the symmetry-breaking dimerization transition [22,36]. In addition, the dielectric constant exhibits two-step changes upon cooling at pressures $P < 9$ kbar coincidentally at the temperatures of the charge transfer and the lattice symmetry breaking determined by the neutron and NQR measurements. These features signify that some change occurs in the system prior to the long-range dimer order; possibly, the charge-transfer transition line is separated from the dimerization transition line on the phase diagram. This bifurcation indicates the existence of a non-dimerized paraelectric ionic (I_{para}) phase between the two lines (charge transfer and dimer order). We note that the above-mentioned a_g-mode activation does not necessarily entail the long-range order of dimerization, as discussed later. Moreover, it was predicted that the charge-transfer transition line has a critical endpoint because the charge transfer transition itself is not accompanied by lattice symmetry breaking [22,36,61,65]. As mentioned in Section 2.2, if the I_{para} phase is stabilized, there appears to be room for the emergence of topological charge transport, which is immobile in the 3D ferroelectric dimer order, in the vicinity of the boundary between the N and I_{para} phases. However, experiments with varying temperatures were only limited to below 9 kbar [22,36,60]. Thus, further investigations at higher pressures are awaited.

3.2. *Revisited Phase Diagram: Ionic Paraelectric Phase Hosting Dimer-Liquid State*

To clarify the existence of the I_{para} phase, we conducted a ^{35}Cl-NQR measurement, which is a microscopic probe of the charge and lattice states, on the polycrystalline TTF-CA under zero magnetic field in an extended pressure range up to 35 kbar [16]. We used Daphne 7373 (for $P < 20$ kbar) and 7474 (for $20 < P < 35$ kbar) oils as the pressure media. The NQR frequency is proportional to the electric-field gradient at the ^{35}Cl nuclear position in a CA molecule, which is determined by the local charge distribution. As demonstrated by earlier NQR experiments, the charge transfer and lattice dimerization are directly signified by the profile of NQR lines [22,36,66,67]; the charge transfer is captured by the shift of the center of gravity of the lines due to the change in the molecular charge, and the lattice dimerization is indicated by line splitting due to the loss of the inversion center on the CA molecule.

At ambient pressure, we observed two ^{35}Cl NQR spectra above T_c, arising from the two inequivalent Cl sites in each CA molecule (all of the CA molecules are crystallographically equivalent in the N phase [37]). Because both spectra show similar behaviors in response to temperature variation, we investigated the NQR line residing on the higher-frequency side. Upon cooling at ambient pressure, a single ^{35}Cl NQR line observed in the N phase splits into two lines below $T_c = 82$ K, accompanied by a jump in the center of gravity of the spectra (see Figure 4a). This indicates that the charge transfer and the dimerization coincide at ambient pressure, as earlier results indicated [66,67]. With increasing pressure at room temperature, however, the NQR line shows a pronounced shift at ~9 kbar but no line splitting up to 35 kbar (see Figure 4b), suggesting that a sharp charge transfer without static lattice dimerization occurs at ~9 kbar. Moreover, upon cooling at a high pressure of 18.2 kbar, we observed a clear line splitting below 258 K, which was not accompanied by an apparent change in the center of gravity of the split lines, as shown in Figure 4c. These results demonstrate that the non-dimerized ionic paraelectric (I_{para}) phase emerges at high temperatures and pressures due to decoupling between the charge transfer and dimer order. The NQR experiments show that, above 8.5 kbar, the dimerization transition temperature (corresponding to the I_{para}-to-I_{ferro} transition temperature) decreases with pressure (see Figure 4d); the I_{para} phase is extended up to 35 kbar, as depicted in Figure 2.

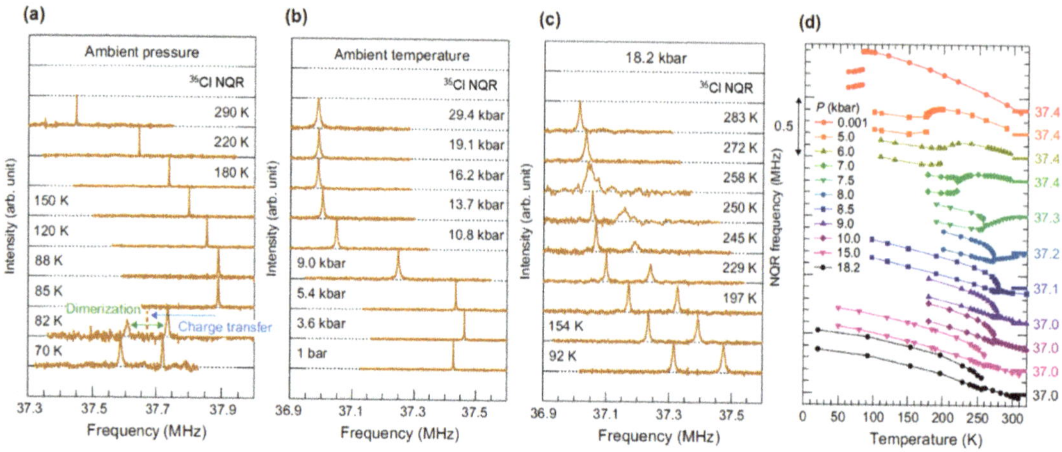

Figure 4. (**a**–**c**) Temperature and pressure profiles of the ^{35}Cl-NQR spectra. (**d**) Temperature dependence of the ^{35}Cl-NQR frequencies at fixed pressures. Reprinted with permission from [16]. Copyright 2018, American Physical Society.

To discuss the profiles of the charge transfer in more detail, we converted the center of gravity of NQR frequency ν_Q (Figure 5a) to the degree of charge transfer ρ using the optical data for reference [33,34]. The details of the estimation of ρ are given in [16]. Here, the main points are briefly summarized below. We assumed that (i) the temperature dependence of ν_Q in the state of $\rho = 0$ obeys the Koukoulas function, which is a well-known formula empirically describing the lattice vibration effect on ν_Q [68] (Figure 6a), (ii) ν_Q depends linearly on ρ at a given temperature (Figure 6a), and (iii) the pressure effect on ν_Q is corrected using the ν_Q data in the neutral CA crystal with $\rho = 0$ (Figure 6b). Based on these assumptions, the relationship between ν_Q and ρ is expressed as $\nu_Q(P, T) = \nu_0 \exp(-\alpha T^2) + A\rho(P, T) + a[1 - \exp(-bP)]$, where A, ν_0, α, a, and b are constants independent of temperature and pressure. The three terms express the three assumptions, respectively. The ρ value estimated using the above formula increases with the application of pressure or decreasing temperature, as shown in Figure 5c, and the jump in ρ at T_c is diminished with increasing pressure up to 7 kbar and becomes quite small at higher pressures (Figure 5d). This appears to be linked to the jump in the line splitting width at T_c, which also decreases with pressure (Figure 5b). These behaviors signify that the NI transition accompanied by the dimer order becomes second-order at high pressures, consistent with previous reports [22,36]. Above 7 kbar, where the charge transfer and dimer order are less coupled, the change in ρ at T_c is nearly zero or takes a negative value, possibly implying that the charge is backflowed by the promotion of molecular orbital hybridization due to dimerization, as suggested in a theoretical study [7].

Is the change from the N to I_{para} phase a phase transition or a crossover? Does the critical endpoint of charge transfer exist? To answer these questions, we plotted the detailed pressure dependence of ρ at fixed temperatures in Figure 5e. The ρ value exhibits a sharp but continuous change around 8 kbar at 280–320 K. There was no discontinuous anomaly in ρ on the boundary between the N and I_{para} phases, indicating that the charge transfer occurs as a crossover. For the critical endpoint of the charge transfer suggested by previous works [22,36,61], it was not captured at least by our experiments. These behaviors are summarized in Figure 5f as the contour plot of ρ on the P-T phase diagram along with the dimerization transition temperature, represented by the white line. This figure clearly indicates that the charge transfer is separated from the dimer order above ~8 kbar, and the charge-transfer crossover line is an extension of the NI transition line determined at pressures below ~7 kbar. Note that the slight jump in ρ at ~7 kbar under 240–260 K in

Figure 5e is an indication of the weak first-order N-to-I$_{ferro}$ transition, not the change from the N to I$_{para}$ phase (see Figure 5f).

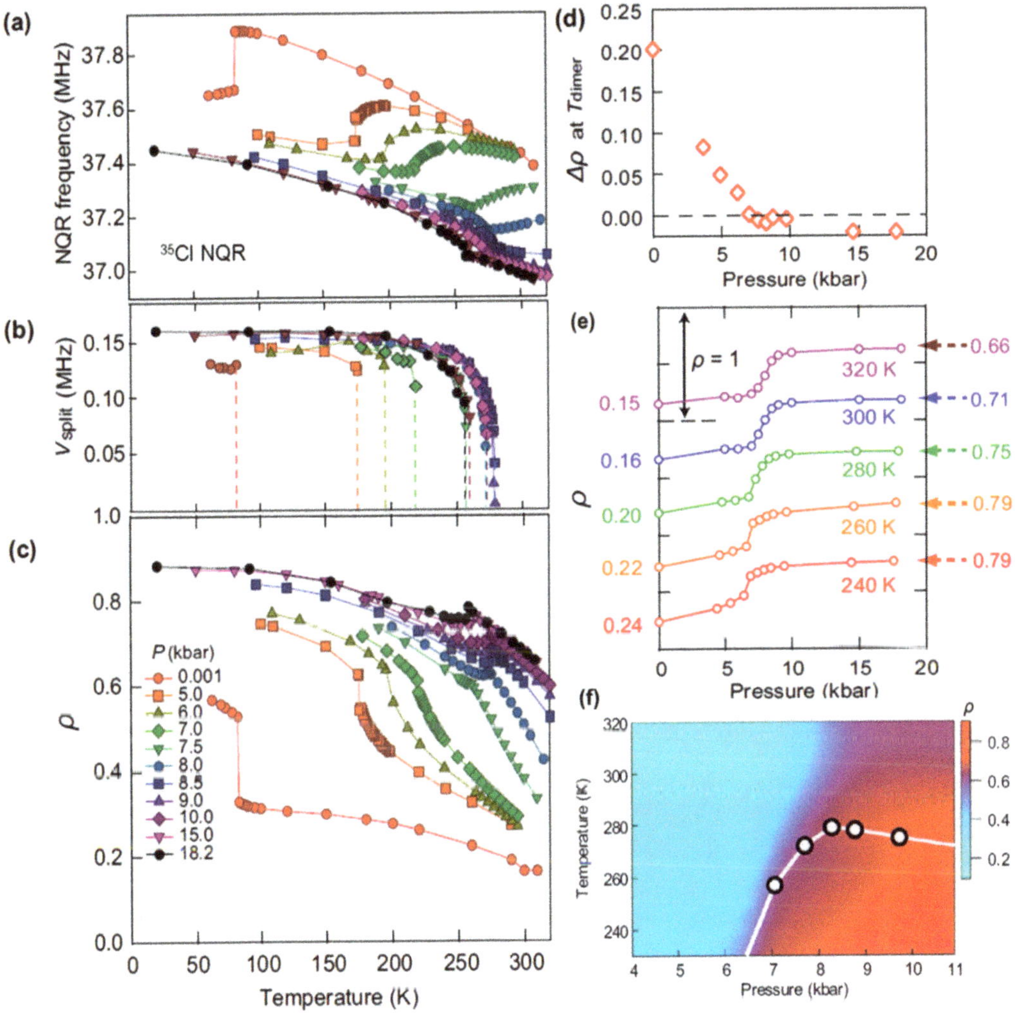

Figure 5. Temperature dependence of the center of gravity of ^{35}Cl-NQR frequency (**a**), line splitting width (**b**), and the degree of charge transfer ρ evaluated from the ^{35}Cl-NQR frequency (**c**) under various pressures. (**d**) The jump in ρ, Δρ, at the dimerization temperature. (**e**) Pressure profiles of ρ at fixed temperatures. (**f**) The contour plot of ρ in the pressure–temperature phase diagram. Open circles represent the phase transition temperatures T_c determined by the NQR experiments, corresponding to those in Figure 2. Reprinted with permission from [16]. Copyright 2018, American Physical Society.

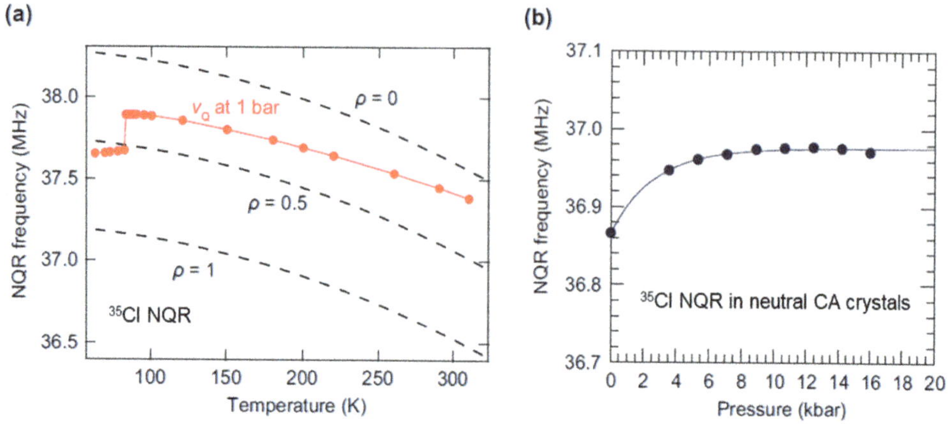

Figure 6. (**a**) Relationship between the ^{35}Cl-NQR frequency ν_Q and the degree of charge transfer ρ at ambient pressure. Red circles are the experimental ν_Q values at 1 bar and red line is a guide for the eye. Broken lines are the calculated temperature dependence of ν_Q in the state of $\rho = 0$, 0.5, and 1, expressed by the Koukoulas function. (**b**) Pressure dependence of the ^{35}Cl-NQR frequency in neutral CA crystals at room temperature. The line is a fit of $a[1 - \exp(-bP)]$ to the data (see text and [16] for details). Reprinted with permission from [16]. Copyright 2018, American Physical Society.

The emergence of the I_{para} phase in the high-temperature and -pressure region appears to be incompatible with the observation of active a_g mode related to the dimerization by infrared spectroscopy above 8 kbar [27,63,64]. This discrepancy can be resolved by the picture that the I_{para} phase is in a liquid state of DA dimers, which are temporally fluctuating at a rate in between the two time scales of the experimental probes, ~10^{-7} s in NQR and ~10^{-12} s in infrared spectroscopy. As further evidence for the dimer liquid, extraordinarily enhanced lattice fluctuations are captured by the ^{35}Cl-NQR relaxation rate $^{35}T_1{}^{-1}$ (see Figure 7). The $^{35}T_1{}^{-1}$ value in the I_{para} phase is two orders of magnitude larger than that in the N phase, suggesting that lattice relaxation is caused by the developed dimer fluctuations in the dimer-liquid state. The slight decrease in the ambient temperature $^{35}T_1{}^{-1}$ with increasing pressures above 10 kbar is possibly related to the decrease in the dimerization transition temperature, which causes a tendency to reduce critical fluctuations at ambient temperature while moving away from the transition temperature at higher pressures.

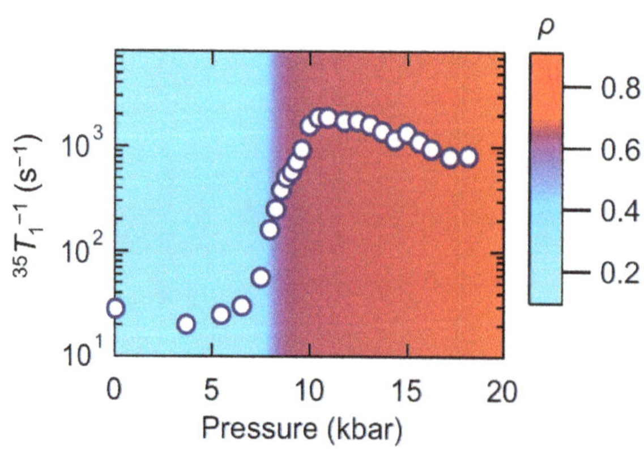

Figure 7. Pressure dependence of the ^{35}Cl-NQR spin–lattice relaxation rate $^{35}T_1{}^{-1}$ at ambient temperature. The color represents the degree of charge transfer ρ and corresponds to the color in Figure 5f. Reprinted with permission from [16]. Copyright 2018, American Physical Society.

In the I_{para} phase hosting the dimer-liquid state, the mobile spin solitons are expected to emerge on the boundaries between the fluctuating opposite polar domains. Near the crossover region between the N and I_{para} phases, the separation of the charge transfer from the long-range dimer order may generate mobile NIDWs in 1D chains, which carry electrical currents, as suggested by theoretical studies [7,9,10].

4. Experimental Evidence for Mobile Topological Excitations Emerging in TTF-CA

4.1. Magnetism in the Ionic Paraelectric Phase: Spin Soliton

In this section, we review the results of NMR experiments on spin excitations in TTF-CA [17]; the aim of this study was to demonstrate the emergence of the mobile spin solitons in the I_{para} phase of TTF-CA. Naively, the I_{para} phase should carry paramagnetic spins on uniform 1D chains; however, the fact that the I_{para} phase hosts the dimer-liquid state implies the emergence of mobile solitonic spins at thermally activated topological boundaries separating oppositely polarized dimer domains.

First, we performed NMR experiments for ^{1}H sites in TTF molecules. Figure 8a shows the pressure profiles of the ^{1}H-NMR spin–lattice relaxation rate $^1T_1{}^{-1}$, probing electron spin fluctuations, at fixed temperatures. At 103 K, where the system shows the N-to-I_{ferro} transition with increasing pressure, $^1T_1{}^{-1}$ is vanishingly small throughout the N and I_{ferro} phases, indicative of nonmagnetic states, which accords with previous ^{1}H-NMR results [69,70] and is reasonable because of the closed-shell structure for the N phase and spin-singlet pairing for the I_{ferro} phase. Remarkably, at higher temperatures, the $^1T_1{}^{-1}$ values are enhanced only in the pressure region above the NI crossover. Consistent behaviors are seen in the temperature dependence of $^1T_1{}^{-1}$ at ambient pressure (the N-to-I_{ferro} transition) and 13 kbar (the I_{para}-to-I_{ferro} transition), as shown in Figure 8b; $^1T_1{}^{-1}$ at ambient pressure is vanishingly small over the whole temperature range below 300 K, whereas $^1T_1{}^{-1}$ at 13 kbar takes large values near room temperature in the I_{para} phase and is drastically suppressed upon entering the I_{ferro} phase. Note that $^1T_1{}^{-1}$ below T_c at 13 kbar adopts a shoulder structure around 200 K; however, the ^{13}C-NMR relaxation rate, $^{13}T_1{}^{-1}$, exhibits no such clear shoulder, although there is a sign of it, as shown later. Although the origin of the different behaviors is not clear, the non-monotonous temperature dependence of the relaxation rate is possibly because two types of spin excitations, such as the spin-triplet and the polaron, are responsible for $T_1{}^{-1}$ in the I_{ferro} phase, as described in Section 4.3; just below T_c, the former, with higher excitation energy, is drastically suppressed upon cooling and almost vanishes around 200 K, whereas the latter, with lower excitation energy, gradually decreases and survives at a lower temperature. The detailed temperature profile of spin excitations upon the I_{para}-to-I_{ferro} transition is discussed in Section 4.3. The contour plot of $^1T_1{}^{-1}$ depicted on the *P*-*T* plane using these data clearly illustrates that the I_{para} phase is the hot spot of activated spins (see Figure 8c). The kink temperatures of $^1T_1{}^{-1}$ well coincide with T_c determined by the NQR measurements described in Section 3.2.

Are these spins activated in the I_{para} phase the mobile spin solitons or alternative spin excitations? To clarify the nature of spin excitations, two experiments were conducted; one is NMR measurement for ^{13}C nuclei on the central double-bonded carbon sites in TTF (Figure 9a), which are more strongly hyperfine-coupled to electron spins than ^{1}H nuclei located on the edges of TTF, and the other is the measurement of the frequency dependence of $^1T_1{}^{-1}$, which probes the frequency profile of the spin dynamics.

In ^{13}C-NMR measurements, the spin susceptibility χ is quantitatively estimated from the NMR spectral shift owing to the strong hyperfine interaction with electron spin. We compare the NMR spectra for the ^{1}H site (Figure 9b) and ^{13}C site (Figure 9c). Each spectrum is composed of multiple peaks due to dipolar interactions between nuclear spins (^{1}H-^{1}H spins for ^{1}H-NMR and ^{13}C-^{13}C spins for ^{13}C-NMR). The ^{13}C-NMR spectrum at 285 K is clearly shifted towards higher frequencies above 7 kbar from its positions in the nonmagnetic N phase at lower pressures or in the nonmagnetic I_{ferro} phase at 14 kbar and 144 K, as shown in Figure 9c. On the other hand, the ^{1}H-NMR spectra at 13 kbar are almost invariant throughout the nonmagnetic I_{para}-to-I_{ferro} transition upon cooling. These indicate that the ^{13}C nuclei are strongly hyperfine-coupled with electron spins, which allows us to evaluate the spin susceptibility. Note that the ^{13}C-NMR spectral shifts observed below 6 kbar of ~82 ppm are attributed to the chemical shift, not contributed to by electron spins, because they fairly well agree with the value observed in the I_{ferro} phase of the nonmagnetic state, represented by the gray line in Figure 9c, which is independent of temperature, as described in Section 4.3.

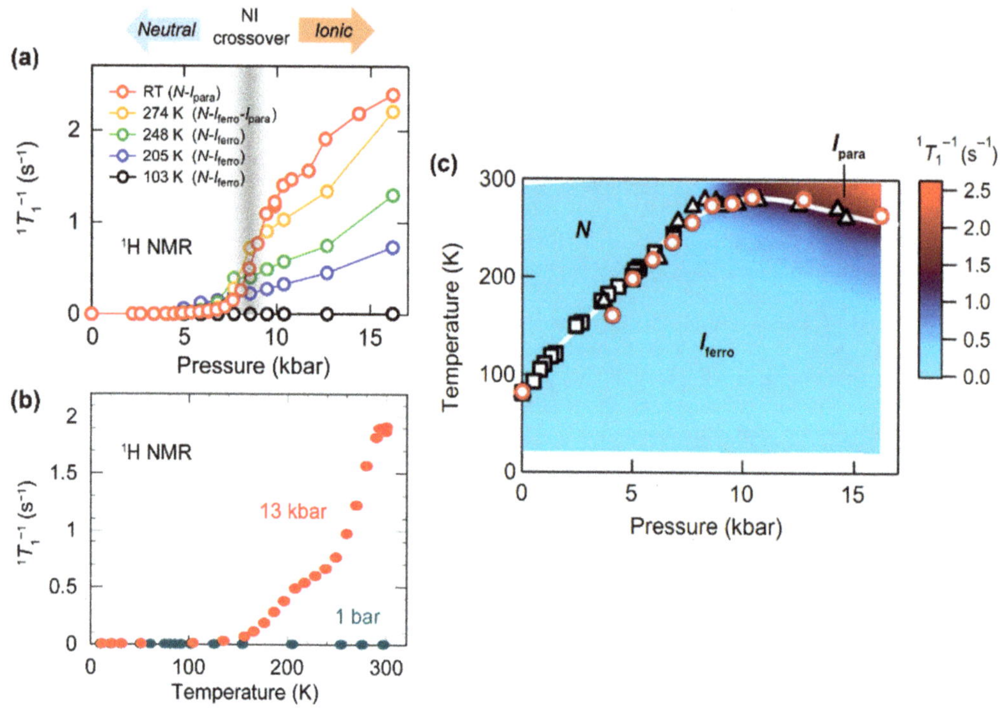

Figure 8. (**a**) Pressure profiles of the ^{1}H-NMR spin–lattice relaxation rate $^1T_1{}^{-1}$ at fixed temperatures. (**b**) Temperature profiles of $^1T_1{}^{-1}$ at ambient pressure (blue circles) and 13 kbar (red circles). (c) Contour plot of $^1T_1{}^{-1}$ in the pressure–temperature phase diagram. Red open circles are the phase transition temperatures T_c determined by the kink temperatures of $^1T_1{}^{-1}$. Black open squares and triangles represent T_c determined by neutron and NQR measurements, respectively [16,36]. Reprinted with permission from [17]. Copyright 2018, The Authors, some rights reserved; exclusive licensee AAAS. Distributed under a CC BY-NC 4.0 License (http://creativecommons.org/licenses/by-nc/4.0/ (accessed on 30 November 2021)).

If the I_{para} phase is not in the dimer-liquid state, the spins should form homogeneous localized spin chains described by the 1D antiferromagnetic Heisenberg model (AFHM). Thus, by comparing the 1D AFHM calculations of χ and the nuclear spin–lattice relaxation rate $^{13}T_1{}^{-1}$ to the experimental χ and $^{13}T_1{}^{-1}$ values, we examine whether the present spin state can be described by the 1D AFHM or not. In the I_{para} phase at 14 kbar, the ^{13}C-NMR spin shift, S, proportional to χ, reaches 51 ppm (Figure 9d), and $^{13}T_1{}^{-1}$ takes a value of 11 s^{-1} (Figure 9e). The spin shift S is expressed as $S = a_{\parallel}\chi/\mu_B N_A$, where N_A is the Avogadro constant, μ_B is the Bohr magneton, and $a_{\parallel}$ is the hyperfine coupling component parallel to the magnetic field H. Through this expression, S is converted to an χ value of 2.6×10^{-5} emu/TTF mol at 14 kbar. Note that we used $^{13}a_{\parallel} = 10.8$ kOe/μ_B of the analogous material, $(TMTTF)_2X$ (X = Br and AsF_6) [71,72], because $^{13}a_{\parallel}$ of TTF-CA is unknown. The hyperfine coupling tensor is mainly determined by the charge density of the HOMO of

TTF, which is concentrated on the central carbons and sulfurs (Figure 9a) [73]. Thus, the difference in the molecular terminal structure between TTF (with hydrogens) and TMTTF (with methyl groups) is not influential on the HOMO density at the central carbon, namely, ${}^{13}a_{\parallel}$.

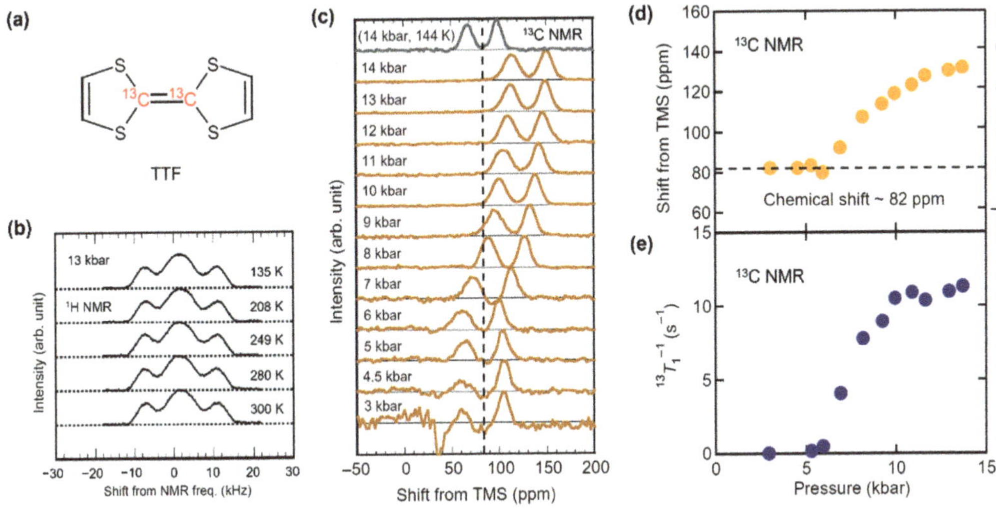

Figure 9. (**a**) ^{13}C-enriched TTF molecule, in which the central double-bonded carbon atoms are substituted by ^{13}C isotopes with a 99% concentration. (**b**) Temperature profile of ^{1}H-NMR spectra at 13 kbar. (**c**) ^{13}C-NMR spectra measured at 285 K under various pressures. The gray spectrum was measured at 14 kbar and 144 K in the ionic ferroelectric phase. The origin of the shift corresponds to the resonance frequency of tetramethylsilane (TMS). Pressure profiles of the ^{13}C-NMR spectral shift (**d**) and spin–lattice relaxation rate ${}^{13}T_1{}^{-1}$ (**e**) at 285 K. Reprinted with permission from [17]. Copyright 2018, The Authors, some rights reserved; exclusive licensee AAAS. Distributed under a CC BY-NC 4.0 License (http://creativecommons.org/licenses/by-nc/4.0/ (accessed on 30 November 2021)).

Then, according to the theoretical calculations of χ by Bonner and Fisher [74] and Estes et al. [75] for the uniform 1D AFHM, the experimental value of $\chi = 2.6 \times 10^{-5}$ emu/TTF mol projects an exchange interaction J of 5900 K, which is unrealistically large. On the other hand, in the scaling theory for the uniform 1D AFHM, the nuclear spin–lattice relaxation rate is expressed as $T_1{}^{-1} = a_{\perp}{}^2 D/\hbar J$, where $a_{\perp}$ is the hyperfine coupling component perpendicular to H, $\hbar$ is the reduced Planck constant, and D is a nonuniversal constant giving the magnitude of the dynamical spin susceptibility [76,77]. Using the values of $J = 5900$ K, $D = 0.15$ determined for the ideal 1D Heisenberg spin system Sr_2CuO_3 [77], and ${}^{13}a_{\perp} = 5.6$ kOe/μ_B for $(TMTTF)_2X$ (X = Br and AsF_6), ${}^{13}T_1{}^{-1}$ for 1D AFHM is calculated to be 1.1 s^{-1}, which is one order of magnitude smaller than the experimental value of ${}^{13}T_1{}^{-1} = 11$ s^{-1} at 14 kbar (Figure 9e). If J is evaluated from this formula, the resultant value of $J = 580$ K projects a χ value one order of magnitude larger than the experimental value. These estimations suggest that the spin state in the I_{para} phase cannot be described by the framework of the uniform 1D AFHM. This consequence, in conjunction with the observed dimer-liquid nature, strongly supports the picture that spins are not on each molecular site but dispersed on the boundaries between the fluctuating dimer domains.

Next, to demonstrate the emergence of mobile spin solitons, we measured the frequency dependence of ${}^{1}T_1{}^{-1}$, which gives the spectral density of local field fluctuations, $S(\omega)$, directly related to the soliton dynamics. We carried out measurements in the frequency range of 14–300 MHz at 300 K under 14 kbar in the I_{para} phase. ${}^{1}T_1{}^{-1}$ shows prominent frequency dependence, as shown in Figure 10; at high frequencies, ${}^{1}T_1{}^{-1}$ obeys the $\omega^{-1/2}$

law expected in the 1D diffusion, whereas it saturates to a constant value upon lowering the frequency. This behavior is well fitted by $T_1^{-1} = k_B T(\chi/N_A\mu_B^2)\gamma_N^2 a_{iso}^2 S_{1D\text{-}3D}(\omega_e)$ + const [78,79], where a_{iso} is the isotropic part of the hyperfine coupling tensor, ω_e is the electron Larmor angular frequency, k_B is the Boltzmann constant, γ_N is the nuclear gyromagnetic ratio, and $S_{1D\text{-}3D}(\omega)$ is $S(\omega)$ in a 1D-3D crossover regime for the weakly coupled 1D systems described by [80]

$$S_{1D-3D}(\omega) = \frac{1}{\sqrt{2D_{\parallel}/\tau_{\perp}}}\left(\frac{1+\sqrt{1+(\omega\tau_{\perp})^2}}{1+(\omega\tau_{\perp})^2}\right)^{1/2}, \quad (1)$$

where $D_{\parallel}$ is the 1D diffusion rate, and $1/\tau_{\perp}$ is the 3D cutoff frequency. Fitting this formula to the experimental data gives $D_{\parallel} = 5.1 \times 10^{11}\ \text{s}^{-1}$ and $1/\tau_{\perp} = 5.6 \times 10^{10}\ \text{s}^{-1}$ with the use of ${}^1a_{iso} = -0.39$ kOe/μ_B of TTF-BDT(Cu) [81]. Note that we neglect the anisotropic part of the hyperfine coupling tensor for simplicity. This result indicates that the spin solitons move diffusively along 1D chains with much slower interchain diffusion. The $D_{\parallel}$ value of the order of $10^{11}\ \text{s}^{-1}$ resides in between the frequency windows of the NQR (10^7 Hz) and infrared (10^{12} Hz) probes, which explains the seeming discrepancy between the NQR and infrared observations and is consistent with the temporally fluctuating dimer-liquid picture.

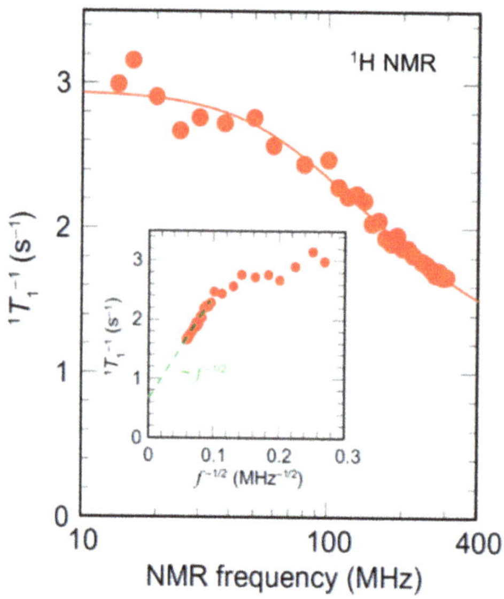

Figure 10. Frequency dependence of ¹H-NMR spin–lattice relaxation rate ${}^1T_1^{-1}$ at 300 K in the ionic paraelectric phase at 14 kbar. Red line is a fit to the data (see text and [17] for details). The inset is a plot of ${}^1T_1^{-1}$ against $f^{-1/2}$. Reprinted with permission from [17]. Copyright 2018, The Authors, some rights reserved; exclusive licensee AAAS. Distributed under a CC BY-NC 4.0 License (http://creativecommons.org/licenses/by-nc/4.0/ (accessed on 30 November 2021)).

Here, the spin-soliton density, n_{ss}, is evaluated from the NMR results. Assuming that the spin solitons behave as Curie spins, $\chi = 2.6 \times 10^{-5}$ emu/TTF mol leads to 1 spin soliton per 25 donor–acceptor (DA) pairs at 285 K under 14 kbar. As another estimation of n_{ss}, if the cutoff time $\tau_{\perp}$ is determined by the collision between the soliton and antisoliton within the 1D chain, we can evaluate the average distance of solitons through the formula of the mean square displacement for the 1D diffusion described by $\langle x(t)^2 \rangle = 2D_{\parallel}t$. Substituting $D_{\parallel} = 5.1 \times 10^{11}\ \text{s}^{-1}$ and $t = \tau_{\perp}$ in this formula, the average distance of spin solitons is estimated to be ~10 DA pairs, which is comparable to $1/n_{ss}$ = 25 DA pairs estimated from the absolute value of χ. Note that the origin of the 3D cutoff has been debatable; there are other conceivable mechanisms, such as the exchange interaction between the spin solitons on adjacent 1D chains or the hopping of solitons to neighbor chains. Thus, the obtained value of $1/n_{ss}$ ~ 10 DA pairs is a lower limit. Anyway, the spin soliton density in the I_{para} phase, 1 spin soliton per 10–25 DA pairs (n_{ss} ~ 0.02–0.05 spins per molecule), is much larger than n_{ss} ~ 10^{-4} spins for frozen solitons in the I_{ferro} phase at ambient pressure [11], signifying that paramagnetic spin solitons are thermally excited only in the I_{para} phase.

During the long history of research on the NI transition since 1981, we successfully provided the first experimental evidence for thermally activated mobile spin solitons. As described in Section 2.2, the steady electrical current carried by the NIDW (or charge solitons) necessitates spin solitons. Thus, the existence of the mobile spin solitons verified by NMR also lends support to anomalous topological charge transport, described below.

4.2. Electrical Conductivity in the Neutral–Ionic Crossover Region: Neutral–Ionic Domain Wall

NMR studies demonstrated the mobile solitonic spin excitations emerging from the ferroelectric order. In the vicinity of the NI transition, another type of topological excitations, namely, NIDWs, can be activated [7–10,52]. The traveling of the two types of topological excitations is expected to lead to unconventional electrical transport. In this section, we review experimental demonstrations of anomalous electrical conduction by NIDWs and spin solitons [18]. Regarding electrical conductivity measurements, some experiments were previously conducted on TTF-CA. Our conductivity measurements differ from the previous ones in the following respects:

(i). We employed the four-terminal method to accurately evaluate the absolute value of the conductivity and the activation energy of charge transport (the two-terminal method was adopted in earlier studies [12,14]).
(ii). The anisotropy of the charge transport was evaluated by measuring the conductivity along the three crystal axes to reveal the nature of unconventional charge carriers.
(iii). We measured the electrical conductivity under finely tuned pressures with a Manganin wire used as a pressure gauge to monitor accurate pressure values near the NI crossover, which allowed us to estimate the activation energy from the conductivity profile along the inclined line parallel to the NI crossover line in the *P*-*T* plane, not at a fixed pressure.

Figure 11a,b show the temperature dependence of the electrical resistivity along the 1D chain (parallel to the *a*-axis) measured with the four-terminal method at fixed pressures. At lower pressures, the resistivity shows an anomaly at T_c and increases exponentially upon cooling, which is compatible with the previously reported two-terminal measurements [12,14]. With increasing pressure up to 9 kbar, the absolute value of resistivity at ambient temperature decreases, and the kink structure at T_c becomes rounded. In the pressure range between 6.6 and 7.5 kbar, the resistivity apparently shows metallic behavior above T_c. At pressures higher than 9 kbar, the resistivity monotonically increases with a slight kink around T_c. Figure 11c shows the contour plot of the *a*-axis conductivity, σ_a, which illustrates a highly conducting state just in the NI crossover region; σ_a reaches a maximum value of 7 S/cm at ~9 kbar (Figure 12a). This value is comparable to the conductivities of organic conductors at room temperature.

The anisotropies of conductivity, σ_a/σ_b and σ_a/σ_c, are also enhanced in association with the peak of σ_a at the same pressure of ~9 kbar (see Figure 12a,b), indicating that the charge carriers resulting in high conductivity in the NI crossover are strongly confined in the 1D chains. Remarkably, this enhancement of conductivity appears between the two insulating phases: the band insulator for the N phase and the ionic Mott insulator for the I phase. These features suggest that the highly conductive charge carriers are associated with unconventional excitations, such as topological NIDW excitations, as previously predicted [12,14,48]. Unbound NIDWs have no way to hop into adjacent chains, resulting in highly 1D transport confined in the chain. Optical measurements indicate that the N and I states dynamically coexist near 9 kbar [34], which accords with the picture that the mobile NIDWs separating the N and I domains are excited in this region. The previously found NIDWs at ambient pressure were in the vicinity of the N-to-I_{ferro} phase transition, and thus, NIDW motion was restricted due to the long-range dimer order. In contrast to this, NIDWs activated near the crossover between the N and I_{para} phase without the dimer order can travel freely to carry the charge current.

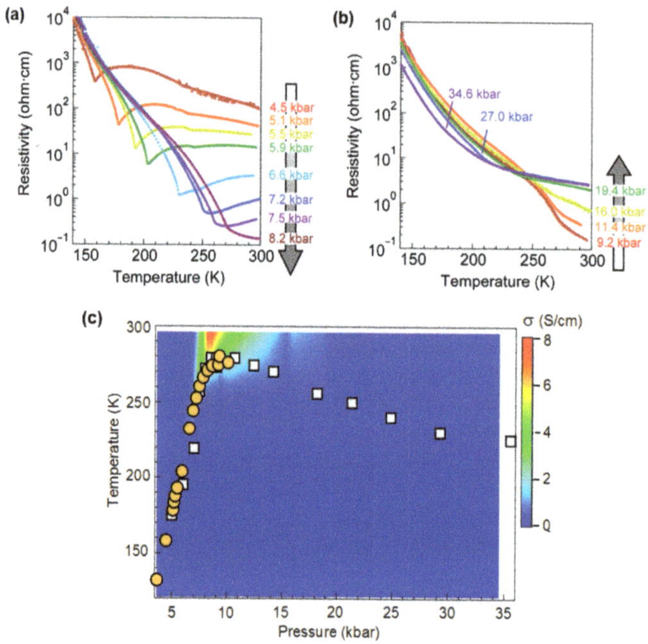

Figure 11. (**a**,**b**) Temperature profiles of the electrical resistivity along the *a*-axis (1D direction) at fixed pressures. (**c**) Contour plot of the electrical conductivity along the *a*-axis σ_a on the pressure–temperature phase diagram. Orange circles represent the phase transition temperatures T_c determined by the kink temperatures of σ_a. Open squares represent T_c determined by NQR measurements [16]. Reprinted with permission from [18]. Copyright 2019, The Authors, some rights reserved; exclusive licensee AAAS. Distributed under a CC BY-NC 4.0 License (http://creativecommons.org/licenses/by-nc/4.0/ (accessed on 15 November 2021)).

To evaluate the activation energy of the NIDW, we performed more detailed resistivity measurements under finely tuned pressures near the NI crossover region. As mentioned above, the temperature profile of resistivity at fixed pressures appears to be of metallic nature near the NI crossover (Figure 11a). This is probably because the NI crossover line denoted by $P_c(T)$, which is determined as the pressure at which the conductivity shows a peak at a fixed temperature, is tilted in the *P*-*T* plane (see Figure 13a), and the activation energy of the NIDW depends on the deviation from the NI crossover line, $\Delta P = P - P_c$. Indeed, the conductivity along inclined lines parallel to the NI crossover line (at $\Delta P = 0$) shows insulating behavior, as shown in Figure 13b. Its Arrhenius plot gives the activation energy E_g of 0.055 eV, which is one order of magnitude smaller than half of the charge-transfer excitation energy, 0.6–0.7 eV [21,82]; it is rather comparable to the theoretically predicted NIDW excitation energy of 0.07–0.08 eV [48]. Moreover, as shown in Figure 13c, the ΔP dependence of E_g reaches a minimum at $\Delta P = 0$, which is also consistent with quantum Monte Carlo simulations [48]. These features suggest that the thermally excited NIDWs dividing the two topologically different domains are responsible for the anomalously high and 1D-confined electrical conductivity.

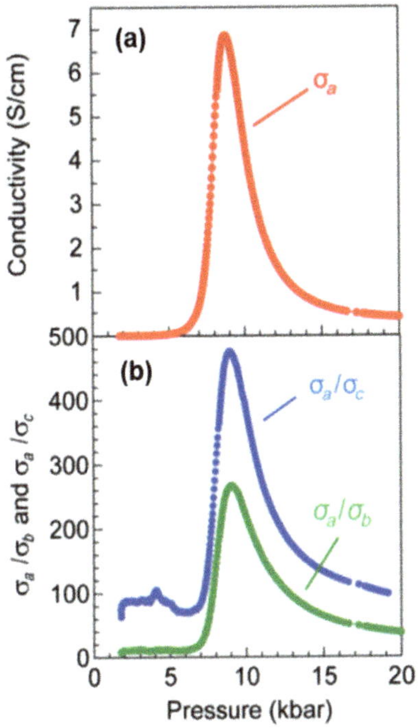

Figure 12. Pressure profiles of electrical conductivity along the *a*-axis σ_a (**a**) and the anisotropies of conductivity, σ_a/σ_b and σ_a/σ_c (**b**), at ambient temperature. Reprinted with permission from [18]. Copyright 2019, The Authors, some rights reserved; exclusive licensee AAAS. Distributed under a CC BY-NC 4.0 License (http://creativecommons.org/licenses/by-nc/4.0/ (accessed on 15 November 2021)).

Here, we estimate the NIDW density, n_{DW}, by mapping the NI transition system to a 1D Ising model [83], which is valid for TTF-CA situated at the NI boundary where the N and I states are degenerate. In this case, the NIDW corresponds to a spinon in the Ising spin system (the N and I states correspond to the up and down spins, respectively). Thus, the correlation length in the Ising model, $\xi \sim 1/\ln(\coth(E_{DW}/2k_BT))$, roughly gives the average distance between the NIDWs; n_{DW} is given by $n_{DW} \sim 1/2\xi \sim \ln(\coth(E_{DW}/2k_BT))/2$. Using $E_{DW} = 0.055$ eV evaluated at $\Delta P = 0$, n_{DW} is estimated at 1 NIDW per ~5 DA pairs on the NI crossover line at ambient temperature. This value is compared to the spin soliton density n_{ss}, 1 spin soliton per 10–25 DA pairs, as described in Section 4.1. Considering that ~10 DA pairs contain one I domain on average, it turns out that a considerable fraction of the I domains contain a spin soliton. This situation is in agreement with the theoretically suggested picture that NIDWs carry the stationary electrical current in cooperation with spin solitons [7,9,10].

Why are spin solitons required for a steady electrical current? We briefly explain the transport mechanism by topological excitations using the schematic illustrations depicted in Figure 14, where the N and I states are represented as D^0A^0 and D^+A^-, respectively, for simplicity.

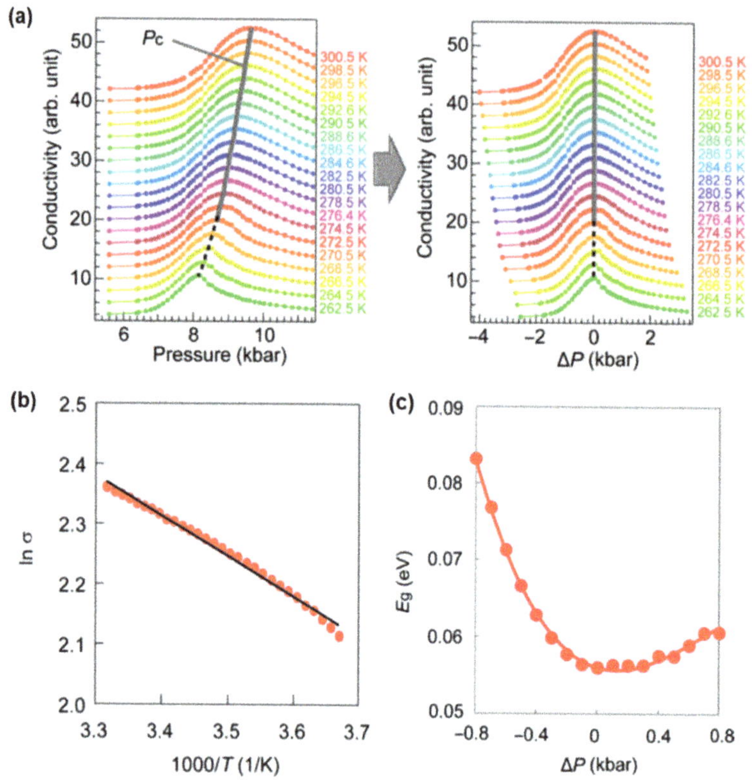

Figure 13. (**a**) Electrical conductivity under finely tuned pressures at fixed temperatures (left panel). Plot of conductivity against $\Delta P = P - P_c$ (right panel). P_c is determined by the peak pressure of conductivity at a given temperature. (**b**) Activation plot of conductivity at $\Delta P = 0$. Black line is a fit of the single-exponential function to conductivity. (**c**) Activation energy, E_g, evaluated at various ΔP. Reprinted with permission from [18]. Copyright 2019, The Authors, some rights reserved; exclusive licensee AAAS. Distributed under a CC BY-NC 4.0 License (http://creativecommons.org/licenses/by-nc/4.0/ (accessed on 15 November 2021)).

Step I: A pair of NIDWs with effective charges, $+\rho_{NIDW}$ and $-\rho_{NIDW}$, is thermally excited in an N domain, where $\pm\rho_{NIDW}$ is the topological charge of $\pm e(\rho_I - \rho_N)/2$.

Step II: An electric field applied to the left separates NIDWs with $+\rho_{NIDW}$ and $-\rho_{NIDW}$ in opposite directions to expand the I_B domain with the D^+A^- pattern.

Step III: A pair of spin solitons with effective charges, $+\rho_{SS}$ and $-\rho_{SS}$, is thermally excited in the I_B domain, where $\pm\rho_{SS}$ is the topological charge of $\pm e(1 - \rho_I)$. It generates an I_A domain with the oppositely polarized A^-D^+ pattern to the I_B domain, which has the D^+A^- pattern.

Step IV: A pair of NIDWs is excited in the I_A domain and grows to an N domain due to an electron transfer to the right.

Consequently, the positive charges, $+\rho_{NIDW}$ and $+\rho_{SS}$, and the negative charges, $-\rho_{NIDW}$ and $-\rho_{SS}$, flow to the left and right, respectively, giving a steady electrical current. If the spin solitons are absent, Step III is replaced with Step III′: allowed excitations of the NIDW pair in the I_B domain are those with $-\rho_{NIDW}$ and $+\rho_{NIDW}$ on the left and right sides, respectively, which are not separated from each other but are recombined under this electric field configuration, resulting in no steady current. In short, the steady current necessitates excitations of oppositely polarized I domains, which are made possible by spin soliton excitations. Thus, the presence of spin solitons is crucial for the generation of steady current, as theoretically proposed [7,9,10].

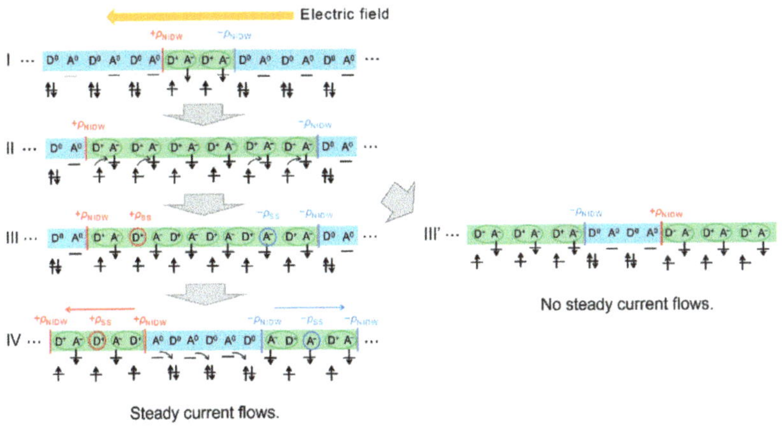

Figure 14. Schematic illustrations of the mechanism of the steady electrical current carried by topological excitations. The neutral and ionic states are represented as D^0A^0 and D^+A^-, respectively, for simplicity. Reprinted with permission from [18]. Copyright 2019, The Authors, some rights reserved; exclusive licensee AAAS. Distributed under a CC BY-NC 4.0 License (http://creativecommons.org/licenses/by-nc/4.0/ (accessed on 15 November 2021)).

Finally, we discuss the electrical conduction carried by topological excitations in the N and I_{para} phases far from the NI crossover. In Figure 15, the pressure dependence of conductivity is shown along with that of ^{13}C-NMR $^{13}T_1{}^{-1}$ to compare the behaviors of spin and charge excitations. For $P < P_c$, $^{13}T_1{}^{-1}$ exhibits an exponential decrease as the system departs from the NI crossover, which is scaled to the conductivity. On the other hand, when the system enters the I_{para} phase for $P > P_c$, the conductivity begins to decline, whereas $^{13}T_1{}^{-1}$ retains large values. These suggest that the charge and spin degrees of freedom are coupled in the N phase, whereas they are decoupled in the I phase. The behavior of the N phase is explainable if the charge carrier is a polaron or equivalently coupled charge and spin solitons with an unfractionalized charge of $\pm e$ and a spin of 1/2, which carries charge and spin inseparably, as depicted in Figure 16. In the I phase away from the NI crossover, the N domain shrinks to form a charge soliton sandwiched by two NIDWs, whose density decreases with pressure. This explains why conductivity decreased when P increased from P_c. Conversely, the spin solitons responsible for the $^{13}T_1{}^{-1}$ behavior are likely to be more excited because their excitation energy is diminished for P higher than P_c, as theoretically suggested [7,9,10]. These decoupled behaviors of charge and spin solitons explain the observation shown in Figure 15. We note that the steady current in the I phase needs both charge and spin solitons, and the conductivity is governed by the density of minor carriers, which are charge solitons. The quantitative estimation of the charge soliton density in the I_{para} phase is described in Section 4.3.

4.3. Binding Transition of Solitons upon Space-Inversion Symmetry-Breaking Ferroelectric Order

The mobile topological excitations mentioned in Sections 4.1 and 4.2 emerge only in the nonordered paraelectric regime; these are not vital in the long-range ordered phase because free solitons, if any, would interrupt the long-ranged order [84]. What form do the spin and charge excitations take in the long-range ordered phase? How do solitons detected in the nonordered paraelectric regime conform to the 3D ferroelectric ordered state? In this section, we review the studies on the profiles of spin and charge excitations upon entering the long-range ferroelectric ordered (I_{ferro}) phase from the nonordered I_{para} phase by resistivity, NMR, and NQR measurements [19], the goal of which is to pursue the fate of spin-soliton–charge-soliton composite matter upon the space-inversion symmetry-breaking ferroelectric order.

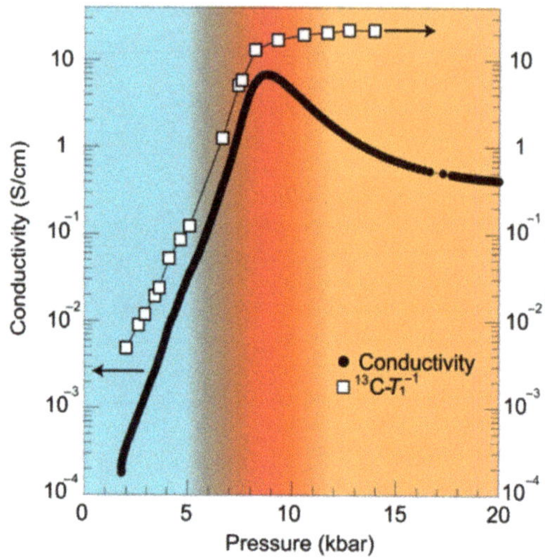

Figure 15. Pressure profiles of the electrical conductivity (closed circles) and ^{13}C-NMR spin–lattice relaxation rate $^{13}T_1{}^{-1}$ (open squares) at ambient temperature. The blue-, red-, and orange-colored regions roughly indicate the neutral phase, NI crossover region, and ionic phase, respectively. Reprinted with permission from [18]. Copyright 2019, The Authors, some rights reserved; exclusive licensee AAAS. Distributed under a CC BY-NC 4.0 License (http://creativecommons.org/licenses/by-nc/4.0/ (accessed on 15 November 2021)).

Before we present the experimental results, to make it clear that the dimer-liquid I_{para} phase of TTF-CA is soliton matter, we evaluated the charge soliton density n_{cs} in the I_{para} phase at 14 kbar using the resistivity data shown in Section 4.2. The electrical transport study strongly suggested that the charge solitons reside in the I_{para} phase because the resistivity in the I_{para} phase is much smaller than that in the N phase (see Figure 15) [18,85]. At 9 kbar in the NI crossover region, the NIDW density, n_{DW}, was estimated at about 1 per 5 DA pairs. As described in Section 4.2, the bound NIDWs sandwiching an N molecule form the charge soliton in the I phase, so $n_{cs} = n_{DW}/2$. At ambient temperature, the σ value at 9 kbar is reduced by an order of magnitude at 14 kbar (see Figure 15), which suggests that n_{cs} is 1 per ~100 DA pairs at 14 kbar, assuming that the pressure dependence of σ depends on only the charge carrier density. On the other hand, the spin-soliton density, n_{ss}, is 1 per 10–25 DA pairs, as deduced from NMR data (see Section 4.1). Thus, the I_{para} phase at 14 kbar hosts soliton matter composed of spin solitons (majority) and charge solitons (minority) with a 10:1–4:1 population (see Figure 16).

First, we performed ^{13}C-NMR measurements at 14 kbar to examine the nature of spin excitations. As shown in Figure 17a,b, both the spin shift, S, and $^{13}T_1{}^{-1}$ exhibit clear kinks at T_c ~ 270 K, below which they decrease upon cooling, consistent with the nonmagnetic state of the I_{ferro} phase. However, the slopes in their Arrhenius plots are different (see Figure 17c); the spin shift multiplied by temperature follows $TS \propto \exp(-\Delta_s/k_BT)$ with a spin excitation gap $\Delta_s$of 3240 K for 200 K < T < T_c, whereas the variation in $^{13}T_1{}^{-1}$ is more gradual in the same temperature range, although the activation energy is not well defined. Here, we discuss the large gap of Δ_s= 3240 K. For conventional spin-Peierls systems, T_c and the singlet–triplet gap Δ hold the relation Δ/k_BT_c ~ 1.76 (the BCS relationship) or 2.47 (obtained with a bosonization method [86]). In the present case, however, Δ_s/k_BT_c yields

~12, which is too large to regard the I_{para}-to-I_{ferro} phase transition as the conventional spin-Peierls transition for 1D Heisenberg spins, supporting the picture that the I_{para} phase hosts mobile spin solitons (see Sections 4.1 and 4.2).

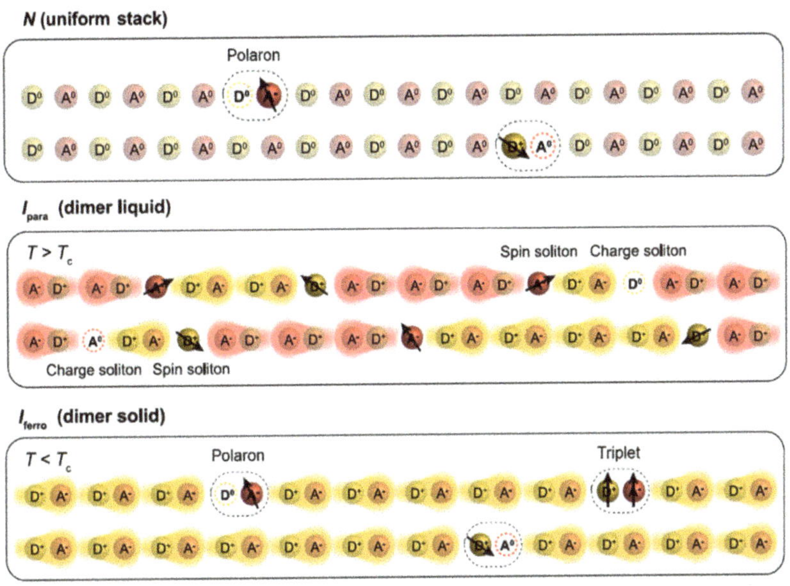

Figure 16. Schematic illustrations of the spin and charge excitations in the neutral (N), ionic paraelectric (I_{para}), and ferroelectric (I_{ferro}) phases. The neutral and ionic states are represented as D^0A^0 and D^+A^- with degrees of charge transfer ρ = 0 and 1, respectively, for simplicity, although ρ takes intermediate values in the neutral and ionic states in reality. The lower two panels are reprinted with permission from [19]. Copyright 2021, American Physical Society.

Next, we discuss the difference between the decreases in the spin shift and ${}^{13}T_1{}^{-1}$ below T_c. For conventional singlet–triplet excitations, the temperature profiles of S and $T_1{}^{-1}$ in the spin-singlet phase should not significantly differ [87]. Thus, the observed behaviors of S and $T_1{}^{-1}$ imply the presence of at least two types of spin excitations below T_c. In the I_{ferro} phase, soliton excitations should be in pairs so as not to violate the 3D ferroelectric order (see Figure 16); one is the triplet spin soliton pair excitation of charge-neutral spin 1, and the other is the spin soliton–charge soliton pair excitation, equivalently polarons, with a charge e and a spin 1/2, as discussed for conducting polymers [3]. The polarons carry not only spins but also charges without violating the ferroelectric dimer order in the I_{ferro} phase. As estimated above, the spin soliton density n_{ss} is larger than the charge soliton density n_{cs} in the I_{para} phase above T_c, suggesting that the magnetism just below T_c should be dominated by bound spin soliton pairs, whose triplet excitations cause a steep decrease in the spin shift. On the other hand, well below T_c, where triplet excitations with a large gap of 3240 K almost vanish, the polaron contribution is expected to be dominant in ${}^{13}T_1{}^{-1}$. These pictures signify that the I_{para}-to-I_{ferro} phase transition is a binding transition of solitons upon the symmetry-breaking dimer order.

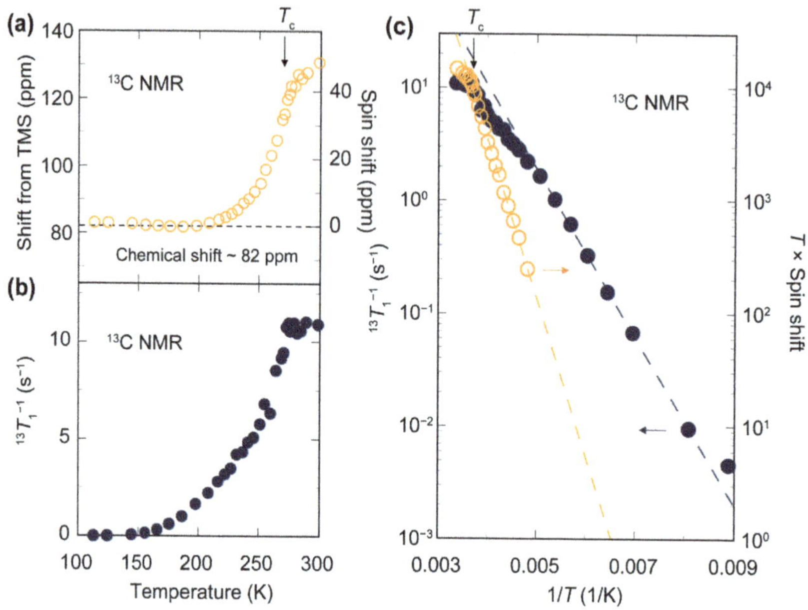

Figure 17. Temperature dependence of ^{13}C-NMR spectral shift (**a**) and spin–lattice relaxation rate $^{13}T_1{}^{-1}$ (**b**) at 14 kbar. (**c**) Activation plots of spin shift multiplied by temperature (orange open circles) and $^{13}T_1{}^{-1}$ (blue closed circles). Reprinted with permission from [19]. Copyright 2021, American Physical Society.

To seek further evidence for the emergence of polarons, we investigated the lattice properties using ^{35}Cl-NQR spectroscopy. The motion of polarons should be followed by lattice distortion. Thus, the ^{35}Cl-NQR spin–lattice relaxation rate $^{35}T_1{}^{-1}$, probing the lattice fluctuations through nuclear quadrupolar interaction, gives valuable insights into the lattice dynamics. $^{35}T_1{}^{-1}$ exhibits a divergent peak at T_c, which is attributable to the critical lattice fluctuations associated with the 3D ferroelectric dimerization transition, and becomes suppressed upon cooling (see Figure 18). Below 140 K, $^{35}T_1{}^{-1}$ roughly obeys the T^2 law, which is the conventional phonon contribution by the two-phonon Raman process [88]. Above 140 K, however, $^{35}T_1{}^{-1}$ is drastically enhanced from the T^2 law, indicating that another relaxation mechanism appears. Remarkably, it is scaled to $^{13}T_1{}^{-1}$ for $140 < T < 250$ K ($< T_c$), signifying that the quadrupolar and magnetic relaxations are attributed to common excitations. This strongly indicates that the spin carriers causing the magnetic relaxation below 250 K are not conventional band quasiparticles but polarons dragging lattice distortion.

Polarons also carry an electrical current. Figure 19a shows the temperature dependence of electrical resistivity at 14 kbar, which exhibits insulating behavior below ambient temperature with a slight kink at T_c. The activation energy of conductivity σ is evaluated to be $\Delta_\sigma \sim 2100$ K in the I_{ferro} phase (see the inset of Figure 19a). If the band quasiparticles are carriers, Δ_σ would be larger than half of the charge-transfer gap of ~0.7 eV (~8100 K) evaluated in optical studies [21,82]. Thus, the substantially lower charge excitation gap suggests that polarons with excitation energies reduced by lattice relaxation are carriers of electrical current in the I_{ferro} phase, distinct from the band quasiparticle excitations.

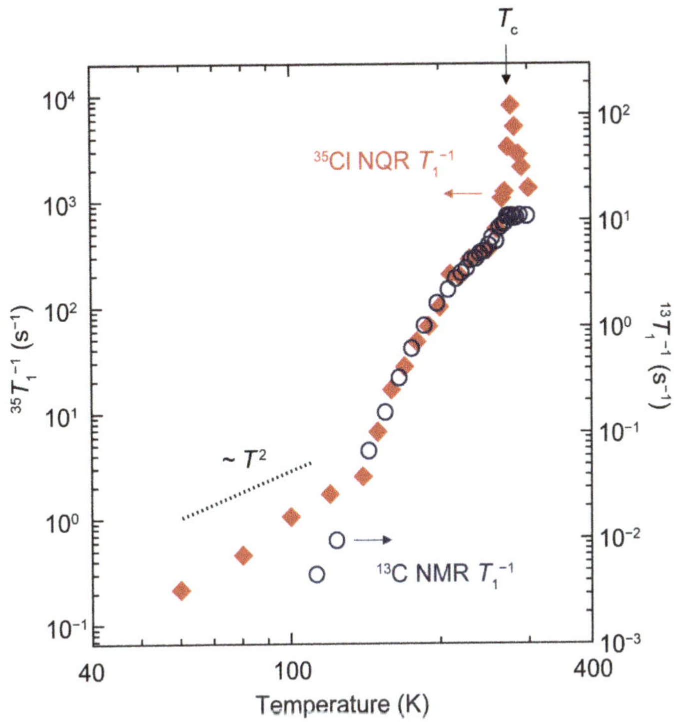

Figure 18. Comparison between the ^{35}Cl-NQR and ^{13}C-NMR spin–lattice relaxation rates, $^{35}T_1{}^{-1}$ (red closed diamonds) and $^{13}T_1{}^{-1}$ (blue open circles), at 14 kbar. Reprinted with permission from [19]. Copyright 2021, American Physical Society.

We demonstrated that there are two types of spin excitations in the I_{ferro} phase: the triplet-type bound spin soliton pairs and the polaronic bound pairs of spin and charge solitons (Figure 16). Finally, we tried to distinguish the two contributions to magnetism. When the polarons move diffusively, the polaron contribution to $^{13}T_1{}^{-1}$, $(^{13}T_1{}^{-1})_p$, is expressed as $(^{13}T_1{}^{-1})_p \propto n_p/\sqrt{aD_\parallel}$, where n_p is the density of the polarons, a is the anisotropy parameter defined by $D_\perp/D_\parallel$, and $D_\parallel$ ($D_\perp$) is the diffusion constant of the polarons parallel (perpendicular) to the 1D chain. On the other hand, σ is expressed as $\sigma = n_p e^2 D_\parallel / k_B T$. By combining the two equations and assuming that n_p obeys the activation form (see [19] for details), the temperature profile of $D_\parallel$ is obtained, as shown in the inset of Figure 19b. Note that we used the anisotropy of conductivity as a [19]. At low temperatures, $D_\parallel$ shows an activation behavior with an energy barrier of 240 K, which is remarkably near the Peierls-coupled optical phonon frequencies of ~120–180 K in the I_{ferro} phase [89]. This suggests that the polarons thermally hop over the energy barrier with the assistance of Peierls phonons. By substituting the obtained $D_\parallel$ value and the activation form of n_p in the above formula of $(^{13}T_1{}^{-1})_p$, we evaluate $(^{13}T_1{}^{-1})_p$ as shown in Figure 19b, which nearly coincides with the experimental $^{13}T_1{}^{-1}$ below 250 K; however, at 250 K < T < T_c, the experimental values deviate upward from $(^{13}T_1{}^{-1})_p$. This additional contribution is very probably the triplet excitations of bound spin soliton pairs with a large gap of Δ_s = 3240 K that appear only just below T_c because the majority are spin solitons in the I_{para} phase above T_c.

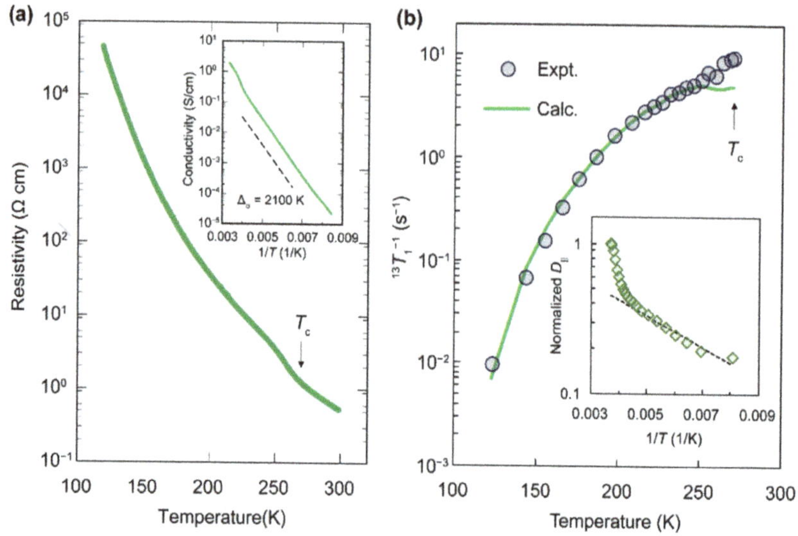

Figure 19. (**a**) Temperature dependence of electrical resistivity at 14 kbar. Inset: Activation plot of electrical conductivity at 14 kbar. (**b**) Comparison between the experimental ${}^{13}T_1{}^{-1}$ and the calculated polaron contribution to ${}^{13}T_1{}^{-1}$, $({}^{13}T_1{}^{-1})_\text{p}$ (see text). The inset shows the normalized 1D diffusion constant. Reprinted with permission from [19]. Copyright 2021, American Physical Society.

The binding transition of solitons has never been observed in a conducting polymer, which does not show a 3D order. Thus, the presented finding of the soliton organization upon the symmetry-breaking transition revealed novel aspects of soliton physics that have ever been studied in one dimension.

5. Summary

In this review, we report recent progress in the study of mobile topological spin and charge excitations in a neutral–ionic (NI) transition system, TTF-CA. The solitons and neutral–ionic domain walls (NIDWs) reported in the past were frozen defects and could not carry an electrical current because these excitations were locked due to the long-range dimer order. In contrast to these observations, we captured mobile topological excitations that are thermally excited in the nonordered paraelectric phase in TTF-CA, as summarized below:

(i). The revisited pressure–temperature phase diagram contains the paraelectric ionic (I_{para}) phase extended in the high-pressure region up to 35 kbar. The N-to-I_{para} phase boundary is a crossover, not a phase transition. The I_{para} phase hosts the dimer-liquid state, providing a chance for the emergence of mobile NIDWs near the N-to-I_{para} crossover and solitons in the I_{para} phase.

(ii). In the I_{para} phase, spin solitons are thermally excited as mobile boundaries dividing fluctuating dimer domains in 1D chains and contribute to the anomalous topological charge transport in cooperation with the NIDW.

(iii). Near the N-to-I_{para} crossover, mobile NIDWs with topological charges carry 1D-confined large electrical conduction in cooperation with spin solitons. This is the first demonstration of the topological charge transport carried by NIDWs and spin solitons in the NI transition system.

(iv). In the 3D ferroelectric ordered (I_{ferro}) phase, spin solitons and charge solitons that are free in the I_{para} phase undergo a binding transition to form two-component composite pairings: neutral spin soliton pairs and polaronic spin-soliton–charge-soliton pairs. The polaronic pairs carry magnetism and electrical conduction with the assistance of Peierls-coupled optical phonons.

The topological charge transport revealed here is not the one that occurs in topological boundaries, as has been extensively studied in topological insulators, but originates from the excitations and motions of the topological boundaries themselves. This novel charge transport is a significant addition to soliton physics, which has been developed mainly for conducting polymers, and is expected to open a novel paradigm for broad transport-related phenomena such as thermoelectric effects.

Author Contributions: K.S. wrote the manuscript with the assistance of K.K., R.T., K.M. and H.O. commented on the manuscript. All authors have read and agreed to the published version of the manuscript.

Funding: This work was supported by the JSPS Grant-in-Aids for Scientific Research (Grant Numbers JP25220709, JP16H06346, JP17K05532, JP17K05846, JP18H05225, JP19H01846, JP20K20894, JP20KK0060, JP21K18144, and JP21H04988); by the Japan Science and Technology Agency, CREST (Grant Number JPMJCR1661); by the Mitsubishi Foundation (Grant Number 202110014); and by the Murata Science Foundation.

Institutional Review Board Statement: Not applicable.

Informed Consent Statement: Not applicable.

Data Availability Statement: Not applicable.

Acknowledgments: We would like to thank F. Iwase, M. Hosoda, T. Nishikawa, A. Katougi, S. Horiuchi, R. Kato, and T. Miyamoto for their experimental collaborations. We also thank H. Fukuyama, N. Nagaosa, and M. Ogata for fruitful discussions.

Conflicts of Interest: The authors declare no conflict of interest.

References

1. Kosterlitz, J.M.; Thouless, D.J. Ordering, metastability and phase transitions in two-dimensional systems. *J. Phys. C Solid State Phys.* **1973**, *6*, 1181–1203. [CrossRef]
2. Thouless, D.J.; Kohmoto, M.; Nightingale, M.P.; Nijs, M. den Quantized Hall conductance in a two dimensional periodic potential. *Phys. Rev. Lett* **1982**, *49*, 405–408. [CrossRef]
3. Heeger, A.J.; Kivelson, S.; Schrieffer, J.R.; Su, W.P. Solitons in conducting polymers. *Rev. Mod. Phys.* **1988**, *60*, 781–850. [CrossRef]
4. Wen, X.G. Topological orders and edge excitations in fractional quantum Hall states. *Adv. Phys.* **1995**, *44*, 405–473. [CrossRef]
5. Brazovskii, S. Ferroelectricity and Charge Ordering in Quasi One-Dimensional Organic Conductors. In *The Physics of Organic Superconductors and Conductors*; Lebed, A., Ed.; Springer: Berlin/Heidelberg, Germany, 2008; pp. 313–355.
6. Hasan, M.Z.; Kane, C.L. Colloquium: Topological insulators. *Rev. Mod. Phys.* **2010**, *82*, 3045–3067. [CrossRef]
7. Nagaosa, N. Theory of neutral-ionic transition in organic crystals. III. Effect of the electron-lattice interaction. *J. Phys. Soc. Jpn.* **1986**, *55*, 2754–2764. [CrossRef]
8. Soos, Z.G.; Painelli, A. Metastable domains and potential energy surfaces in organic charge-transfer salts with neutral-ionic phase transitions. *Phys. Rev. B* **2007**, *75*, 155119. [CrossRef]
9. Fukuyama, H.; Ogata, M. Solitons in the crossover between band insulator and Mott insulator: Application to TTF-Chloranil under pressure. *J. Phys. Soc. Jpn.* **2016**, *85*, 023702. [CrossRef]
10. Tsuchiizu, M.; Yoshioka, H.; Seo, H. Phase competition, solitons, and domain walls in neutral-ionic transition systems. *J. Phys. Soc. Jpn.* **2016**, *85*, 104705. [CrossRef]
11. Mitani, T.; Saito, G.; Tokura, Y.; Koda, T. Soliton formation at the neutral-to-ionic phase transition in the mixed-stack charge-transfer crystal tetrathiafulvalene-*p*-chloranil. *Phys. Rev. Lett.* **1984**, *53*, 842–845. [CrossRef]
12. Mitani, T.; Kaneko, Y.; Tanuma, S.; Tokura, Y.; Koda, T.; Saito, G. Electric conductivity and phase diagram of a mixed-stack charge-transfer crystal: Tetrathiafulvalene-*p*-chloranil. *Phys. Rev. B* **1987**, *35*, 427–429. [CrossRef] [PubMed]
13. Tokura, Y.; Okamoto, H.; Koda, T.; Mitani, T.; Saito, G. Nonlinear electric transport and switching phenomenon in the mixed-stack charge-transfer crystal tetrathiafulvalene-*p*-chloranil. *Phys. Rev. B* **1988**, *38*, 2215–2218. [CrossRef] [PubMed]
14. Okamoto, H.; Komatsu, T.; Iwasa, Y.; Koda, T.; Tokura, Y.; Koshihara, S.; Mitani, T.; Saito, G. Dynamical aspects of neutral-ionic phase transition in organic charge-transfer complex crystals. *Synth. Met.* **1988**, *27*, 189–196. [CrossRef]
15. Okamoto, H.; Mitani, T.; Tokura, Y.; Koshihara, S.; Komatsu, T.; Iwasa, Y.; Koda, T.; Saito, G. Anomalous dielectric response in tetrathiafulvalene-*p*-chloranil as observed in temperature- and pressure-induced neutral-to-ionic phase transition. *Phys. Rev. B* **1991**, *43*, 8224–8232. [CrossRef] [PubMed]
16. Takehara, R.; Sunami, K.; Iwase, F.; Hosoda, M.; Miyagawa, K.; Miyamoto, T.; Okamoto, H.; Kanoda, K. Revisited phase diagram on charge instability and lattice symmetry breaking in the organic ferroelectric TTF-QCl_4. *Phys. Rev. B* **2018**, *98*, 054103. [CrossRef]

17. Sunami, K.; Nishikawa, T.; Miyagawa, K.; Horiuchi, S.; Kato, R.; Miyamoto, T.; Okamoto, H.; Kanoda, K. Evidence for solitonic spin excitations from a charge-lattice-coupled ferroelectric order. *Sci. Adv.* **2018**, *4*, eaau7725. [CrossRef]
18. Takehara, R.; Sunami, K.; Miyagawa, K.; Miyamoto, T.; Okamoto, H.; Horiuchi, S.; Kato, R.; Kanoda, K. Topological charge transport by mobile dielectric-ferroelectric domain walls. *Sci. Adv.* **2019**, *5*, eaax8720. [CrossRef]
19. Sunami, K.; Takehara, R.; Katougi, A.; Miyagawa, K.; Horiuchi, S.; Kato, R.; Miyamoto, T.; Okamoto, H.; Kanoda, K. Fate of soliton matter upon symmetry-breaking ferroelectric order. *Phys. Rev. B* **2021**, *103*, 134112. [CrossRef]
20. McConnell, H.M.; Hoffman, B.M.; Metzger, R.M. Charge transfer in molecular crystals. *Proc. Natl. Acad. Sci. USA* **1965**, *53*, 46–50. [CrossRef]
21. Torrance, J.B.; Vazquez, J.E.; Mayerle, J.J.; Lee, V.Y. Discovery of a neutral-to-ionic phase transition in organic materials. *Phys. Rev. Lett.* **1981**, *46*, 253–257. [CrossRef]
22. Buron-Le Cointe, M.; Collet, E.; Toudic, B.; Czarnecki, P.; Cailleau, H. Back to the structural and dynamical properties of neutral-ionic phase transitions. *Crystals* **2017**, *7*, 285. [CrossRef]
23. Torrance, J.B.; Girlando, A.; Mayerle, J.J.; Crowley, J.I.; Lee, V.Y.; Batail, P.; LaPlaca, S.J. Anomalous nature of neutral-to-ionic phase transition in tetrathiafulvalene-chloranil. *Phys. Rev. Lett.* **1981**, *47*, 1747–1750. [CrossRef]
24. Tokura, Y.; Koda, T.; Mitani, T.; Saito, G. Neutral-to-ionic transition in tetrathiafulvalene-*p*-chloranil as investigated by optical reflection spectra. *Solid State Commun.* **1982**, *43*, 757–760. [CrossRef]
25. Girlando, A.; Marzola, F.; Pecile, C.; Torrance, J.B. Vibrational spectroscopy of mixed stack organic semiconductors: Neutral and ionic phases of tetrathiafulvalene–chloranil (TTF–CA) charge transfer complex. *J. Chem. Phys.* **1983**, *79*, 1075–1085. [CrossRef]
26. Dressel, M.; Peterseim, T. Infrared investigations of the neutral-ionic phase transition in TTF-CA and its dynamics. *Crystals* **2017**, *7*, 17. [CrossRef]
27. Masino, M.; Castagnetti, N.; Girlando, A. Phenomenology of the neutral-ionic valence instability in mixed stack charge-transfer crystals. *Crystals* **2017**, *7*, 108. [CrossRef]
28. Kagawa, F.; Horiuchi, S.; Tokura, Y. Quantum phenomena emerging near a ferroelectric critical point in a donor–acceptor organic charge-transfer complex. *Crystals* **2017**, *7*, 106. [CrossRef]
29. Morimoto, T.; Miyamoto, T.; Okamoto, H. Ultrafast electron and molecular dynamics in photoinduced and electric-field-induced neutral–ionic transitions. *Crystals* **2017**, *7*, 132. [CrossRef]
30. D'Avino, G.; Painelli, A.; Soos, Z.G. Modeling the neutral-ionic transition with correlated electrons coupled to soft lattices and molecules. *Crystals* **2017**, *7*, 144. [CrossRef]
31. Tokura, Y.; Okamoto, H.; Koda, T.; Mitani, T.; Saito, G. Pressure-induced neutral-to-ionic phase transition in TTF-*p*-chloranil studied by infrared vibrational spectroscopy. *Solid State Commun.* **1986**, *57*, 607–610. [CrossRef]
32. Girlando, A.; Pecile, C.; Brillante, A.; Syassen, K. Neutral-ionic interface in mixed stack charge transfer compounds: Pressure induced ionic phase of tetrathiafulvalene-chloranil (TTF-CA). *Solid State Commun.* **1986**, *57*, 891–896. [CrossRef]
33. Horiuchi, S.; Okimoto, Y.; Kumai, R.; Tokura, Y. Anomalous valence fluctuation near a ferroelectric transition in an organic charge-transfer complex. *J. Phys. Soc. Jpn.* **2000**, *69*, 1302–1305. [CrossRef]
34. Matsuzaki, H.; Takamatsu, H.; Kishida, H.; Okamoto, H. Valence fluctuation and domain-wall dynamics in pressure-induced neutral-to-ionic phase transition of organic charge-transfer crystal. *J. Phys. Soc. Jpn.* **2005**, *74*, 2925–2928. [CrossRef]
35. Dengl, A.; Beyer, R.; Peterseim, T.; Ivek, T.; Untereiner, G.; Dressel, M. Evolution of ferroelectricity in tetrathiafulvalene-p-chloranil as a function of pressure and temperature. *J. Chem. Phys.* **2014**, *140*, 244511. [CrossRef]
36. Lemée-Cailleau, M.H.; Le Cointe, M.; Cailleau, H.; Luty, T.; Moussa, F.; Roos, J.; Brinkmann, D.; Toudic, B.; Ayache, C.; Karl, N. Thermodynamics of the neutral-to-ionic transition as condensation and crystallization of charge-transfer excitations. *Phys. Rev. Lett.* **1997**, *79*, 1690–1693. [CrossRef]
37. Le Cointe, M.; Lemée-Cailleau, M.H.; Cailleau, H.; Toudic, B.; Toupet, L.; Heger, G.; Moussa, F.; Schweiss, P.; Kraft, K.H.; Karl, N. Symmetry breaking and structural changes at the neutral-to-ionic transition in tetrathiafulvalene-*p*-chloranil. *Phys. Rev. B* **1995**, *51*, 3374–3386. [CrossRef]
38. Kanai, Y.; Tani, M.; Kagoshima, S.; Tokura, Y.; Koda, T. X-ray evidence for the molecular dimerization in TTF-chloranil. *Synth. Met.* **1984**, *10*, 157–160. [CrossRef]
39. Tokura, Y.; Kaneko, Y.; Okamoto, H.; Tanuma, S.; Koda, T.; Mitani, T.; Saito, G. Spectroscopic study of the neutral-to-ionic phase transition in TTF-Chloranil. *Mol. Cryst. Liq. Cryst.* **1985**, *125*, 71–80. [CrossRef]
40. Kobayashi, K.; Horiuchi, S.; Kumai, R.; Kagawa, F.; Murakami, Y.; Tokura, Y. Electronic ferroelectricity in a molecular crystal with large polarization directing antiparallel to ionic displacement. *Phys. Rev. Lett.* **2012**, *108*, 237601. [CrossRef]
41. Giovannetti, G.; Kumar, S.; Stroppa, A.; Van Den Brink, J.; Picozzi, S. Multiferroicity in TTF-CA organic molecular crystals predicted through Ab initio calculations. *Phys. Rev. Lett.* **2009**, *103*, 266401. [CrossRef]
42. Ishibashi, S.; Terakura, K. Exotic ferroelectricity in tetrathiafulvalene-*p*-chloranil: Anomalous effective charges and a picture in the framework of maximally localized wannier orbitals. *J. Phys. Soc. Jpn.* **2014**, *83*, 073702. [CrossRef]
43. Borisov, V.; Biswas, S.; Li, Y.; Valentí, R. Microscopic modeling of correlated systems under pressure: Representative examples. *Phys. Status Solidi B* **2019**, *256*, 1900229. [CrossRef]
44. Resta, R.; Vanderbilt, D. Theory of Polarization: A Modern Approach. In *Physics of Ferroelectrics: A Modern Perspective*; Rabe, K.M., Ahn, C.H., Triscone, J.-M., Eds.; Springer: Berlin/Heidelberg, Germany, 2007; pp. 31–68.
45. King-Smith, R.D.; Vanderbilt, D. Theory of polarization of crystalline solids. *Phys. Rev. B* **1993**, *47*, 1651–1654. [CrossRef]

46. Vanderbilt, D.; King-Smith, R.D. Electric polarization as a bulk quantity and its relation to surface charge. *Phys. Rev. B* **1993**, *48*, 4442–4455. [CrossRef] [PubMed]
47. Resta, R. Macroscopic polarization in crystalline dielectrics: The geometric phase approach. *Rev. Mod. Phys.* **1994**, *66*, 899–915. [CrossRef]
48. Nagaosa, N.; Takimoto, J. Theory of neutral-ionic transition in organic crystals. II. Effect of the intersite Coulomb interaction. *J. Phys. Soc. Jpn.* **1986**, *55*, 2745–2753. [CrossRef]
49. Nagaosa, N. Theory of neutral-ionic transition in organic crystals. IV. Phenomenological viewpoint. *J. Phys. Soc. Jpn.* **1986**, *55*, 3488–3497. [CrossRef]
50. Kagawa, F.; Horiuchi, S.; Matsui, H.; Kumai, R.; Onose, Y.; Hasegawa, T.; Tokura, Y. Electric-field control of solitons in a ferroelectric organic charge-transfer salt. *Phys. Rev. Lett.* **2010**, *104*, 227602. [CrossRef]
51. Kishida, H.; Takamatsu, H.; Fujinuma, K.; Okamoto, H. Ferroelectric nature and real-space observations of domain motions in the organic charge-transfer compound tetrathiafulvalene-*p*-chloranil. *Phys. Rev. B* **2009**, *80*, 205201. [CrossRef]
52. Buron-Le Cointe, M.; Lemée-Cailleau, M.H.; Cailleau, H.; Ravy, S.; Bérar, J.F.; Rouzière, S.; Elkaïm, E.; Collet, E. One-dimensional fluctuating nanodomains in the charge-transfer molecular system TTF-CA and their first-order crystallization. *Phys. Rev. Lett.* **2006**, *96*, 205503. [CrossRef]
53. Morimoto, T.; Miyamoto, T.; Yamakawa, H.; Terashige, T.; Ono, T.; Kida, N.; Okamoto, H. Terahertz-Field-Induced Large Macroscopic Polarization and Domain-Wall Dynamics in an Organic Molecular Dielectric. *Phys. Rev. Lett.* **2017**, *118*, 107602. [CrossRef] [PubMed]
54. Koshihara, S.; Tokura, Y.; Mitani, T.; Saito, G.; Koda, T. Photoinduced valence instability in the organic molecular compound tetrathiafulvalene-*p*-chloranil (TTF-CA). *Phys. Rev. B* **1990**, *42*, 6853. [CrossRef] [PubMed]
55. Iwai, S.; Okamoto, H. Ultrafast phase control in one-dimensional correlated electron systems. *J. Phys. Soc. Jpn.* **2006**, *75*, 011007. [CrossRef]
56. Miyashita, N.; Kuwabara, M.; Yonemitsu, K. Electronic and lattice dynamics in the photoinduced ionic-to-neutral phase transition in a one-dimensional extended Peierls–Hubbard model. *J. Phys. Soc. Jpn.* **2003**, *72*, 2282–2290. [CrossRef]
57. Ohmura, S.; Mase, T.; Takahashi, A. Terahertz pulse induced transitions between ionic and neutral phases and electronic polarization reversal in TTF-CA. *Phys. Rev. B* **2019**, *100*, 035116. [CrossRef]
58. Watanabe, Y.; Ando, H.; Takahashi, A.; Tomita, N. Nonadiabatic quantum fluctuations in the neutral ground state of tetrathiafulvalene-*p*-chloranil. *Phys. Rev. B* **2019**, *100*, 205205. [CrossRef]
59. Miyamoto, T.; Yada, H.; Yamakawa, H.; Okamoto, H. Ultrafast modulation of polarization amplitude by terahertz fields in electronic-type organic ferroelectrics. *Nat. Commun.* **2013**, *4*, 2586. [CrossRef]
60. Takaoka, K.; Kaneko, Y.; Okamoto, H.; Tokura, Y.; Koda, T.; Mitani, T.; Saito, G. Infrared molecular-vibration spectra of tetrathiafulvalene-chloranil crystal at low temperature and high pressure. *Phys. Rev. B* **1987**, *36*, 3884–3887. [CrossRef]
61. Luty, T.; Cailleau, H.; Koshihara, S.; Collet, E.; Takesada, M.; Lemée-Cailleau, M.H.; Cointe, M.B.-L.; Nagaosa, N.; Tokura, Y.; Zienkiewicz, E.; et al. Static and dynamic order of cooperative multi-electron transfer. *Europhys. Lett.* **2002**, *59*, 619–625. [CrossRef]
62. Metzger, R.M.; Torrance, J.B. Role of the madelung energy in the neutral-ionic phase transition of Tetrathiafulvalene Chloranil. *J. Am. Chem. Soc.* **1985**, *107*, 117–121. [CrossRef]
63. Okamoto, H.; Koda, T.; Tokura, Y.; Mitani, T.; Saito, G. Pressure-induced neutral-to-ionic phase transition in organic charge-transfer crystals of tetrathiafulvalene-*p*-benzoquinone derivatives. *Phys. Rev. B* **1989**, *39*, 10693–10701. [CrossRef] [PubMed]
64. Masino, M.; Girlando, A.; Brillante, A. Intermediate regime in pressure-induced neutral-ionic transition in tetrathiafulvalene-chloranil. *Phys. Rev. B* **2007**, *76*, 064114. [CrossRef]
65. Kishine, J.; Luty, T.; Yonemitsu, K. Ferroelectric phase transition, ionicity condensation, and multicriticality in charge-transfer organic complexes. *Phys. Rev. B* **2004**, *69*, 075115. [CrossRef]
66. Gourdji, M.; Guibé, L.; Péneau, A.; Gallier, J.; Toudic, B.; Cailleau, H. ^{35}Cl NQR observation of the neutral-to-ionic phase transition in tetrathiafulvalene-*p*-chloranil. *Solid State Commun.* **1991**, *77*, 609–612. [CrossRef]
67. Gallier, J.; Toudic, B.; Delugeard, Y.; Cailleau, H.; Gourdji, M.; Peneau, A.; Guibe, L. Chlorine-nuclear-quadrupole-resonance study of the neutral-to-ionic transition in tetrathiafulvalene-chloranil. *Phys. Rev. B* **1993**, *47*, 11688. [CrossRef]
68. Koukoulas, A.A.; Whitehead, M.A. Observations in nuclear quadrupole resonance frequency temperature dependence. *Chem. Phys. Lett.* **1990**, *167*, 379–382. [CrossRef]
69. Yoshinari, Y.; Maniwa, Y.; Takahashi, T.; Mizoguchi, K.; Mitani, T. ^{1}H-NMR studies of neutral-ionic transition in TTF-p-Chloranil. *Synth. Met.* **1987**, *19*, 521–526. [CrossRef]
70. Toudic, B.; Gallier, J.; Boumaza, M.; Cailleau, H. Proton spin-lattice relaxation study of the neutral-to-ionic transition in TTF-chloranil. *J. Phys. Fr.* **1990**, *51*, 1671–1678. [CrossRef]
71. Barthel, E.; Quiron, G.; Wzietek, P.; Jerome, D.; Christensen, J.; Bechgaard, K. NMR in commensurate and incommensurate spin density waves. *Europhys. Lett.* **1993**, *21*, 87–92. [CrossRef]
72. Fujiyama, S.; Nakamura, T. Redistribution of electronic charges in spin-Peierls state in $(TMTTF)_2AsF_6$ observed by ^{13}C NMR. *J. Phys. Soc. Jpn.* **2006**, *75*, 014705. [CrossRef]
73. Katan, C. First-principles study of the structures and vibrational frequencies for Tetrathiafulvalene TTF and TTF-d_4 in different oxidation states. *J. Phys. Chem. A* **1999**, *103*, 1407–1413. [CrossRef]
74. Bonner, J.C.; Fisher, M.E. Linear magnetic chains with anisotropic coupling. *Phys. Rev.* **1964**, *135*, A640. [CrossRef]

75. Estes, W.E.; Gavel, D.P.; Hatfield, W.E.; Hodgson, D.J. Magnetic and structural characterization of dibromo- and dichloro-bis(thiazole)copper(II). *Inorg. Chem.* **1978**, *17*, 1415–1421. [CrossRef]
76. Sachdev, S. NMR relaxation in half-integer antiferromagnetic spin chains. *Phys. Rev. B* **1994**, *50*, 13006–13008. [CrossRef]
77. Takigawa, M.; Motoyama, N.; Eisaki, H.; Uchida, S. Dynamics in the $S = 1/2$ one-dimensional antiferromagnet Sr_2CuO_3 via ^{63}Cu NMR. *Phys. Rev. Lett.* **1996**, *76*, 4612–4615. [CrossRef]
78. Devreux, F. Nuclear relaxation in one-dimensional Hubbard systems. *Phys. Rev. B* **1976**, *13*, 4651–4657. [CrossRef]
79. Nechtschein, M.; Devreux, F.; Greene, R.L.; Clarke, T.C.; Street, G.B. One-dimensional spin diffusion in polyacetylene, $(CH)_x$. *Phys. Rev. Lett.* **1980**, *44*, 356–359. [CrossRef]
80. Mizoguchi, K.; Kume, K.; Shirakawa, H. Frequency dependence of electron spin-lattice relaxation rate at 5–450 MHz in pristine trans-polyacetylene—New evidence of one dimensional diffusive motion of electron spin (neutral soliton) —. *Solid State Commun.* **1984**, *50*, 213–218. [CrossRef]
81. Devreux, F.; Jeandey, C.; Nechtschein, M.; Fabre, J.M.; Giral, L. Electron-proton couplings and local susceptibilities in TTF and TCNQ salts. *J. Phys.* **1979**, *40*, 671–677. [CrossRef]
82. Okamoto, H.; Ishige, Y.; Tanaka, S.; Kishida, H.; Iwai, S.; Tokura, Y. Photoinduced phase transition in tetrathiafulvalene-*p*-chloranil observed in femtosecond reflection spectroscopy. *Phys. Rev. B* **2004**, *70*, 165202. [CrossRef]
83. Bruinsma, R.; Per Bak, J.B. Torrance Neutral-ionic transitions in organic mixed-stack compounds. *Phys. Rev. B* **1983**, *27*, 456–466. [CrossRef]
84. Karpov, P.; Brazovskii, S. Phase transitions in ensembles of solitons induced by an optical pumping or a strong electric field. *Phys. Rev. B* **2016**, *94*, 125108. [CrossRef]
85. Takehara, R.; Miyagawa, K.; Kanoda, K.; Miyamoto, T.; Matsuzaki, H.; Okamoto, H.; Taniguchi, H.; Matsubayashi, K.; Uwatoko, Y. Electron transport in TTF-CA under high pressures. *Physica B* **2015**, *460*, 83–87. [CrossRef]
86. Orignac, E.; Chitra, R. Mean-field theory of the spin-Peierls transition. *Phys. Rev. B* **2004**, *70*, 214436. [CrossRef]
87. Itoh, Y.; Yasuoka, H. Interrelation between dynamical and static spin gaps in quantum spin systems. *J. Phys. Soc. Jpn.* **1997**, *66*, 334–336. [CrossRef]
88. Abragam, A. *The Principles of Nuclear Magnetism*; Oxford University Press: Oxford, UK, 1961.
89. Masino, M.; Girlando, A.; Brillante, A.; Della Valle, R.G.; Venuti, E.; Drichko, N.; Dressel, M. Lattice dynamics of TTF-CA across the neutral-ionic transition. *Chem. Phys.* **2006**, *325*, 71–77. [CrossRef]

symmetry MDPI

Review

Tracking Defects of Electronic Crystals by Coherent X-ray Diffraction

David Le Bolloc'h [1,*], Ewen Bellec [2], Natacha Kirova [1,*] and Vincent L. R. Jacques [1]

1 Laboratoire de Physique des Solides, CNRS, Université Paris-Sud, UMR 8502, 91405 Orsay, France; vincent.jacques@universite-paris-saclay.fr
2 European Synchrotron Radiation Facility, Avenue des Martyrs, 71, CEDEX 9, 38043 Grenoble, France
* Correspondence: david.le-bolloch@universite-paris-saclay.fr (D.L.B.); natacha.kirova@universite-paris-saclay.fr (N.K.)

Abstract: In this article, we review different studies based on advanced X-ray diffraction techniques—especially coherent X-ray diffraction—that allowed us to reveal the behaviour of such symmetry-breaking systems as Charge Density Wave (CDW) and Spin density Wave (SDW), through their local phase. After a brief introduction on the added value of using coherent X-rays, we show how the method can be applied to CDW and SDW systems, in both static and dynamical regimes. The approach allowed us to probe the particular sliding state of CDWs systems by observing them through their phase fluctuations, to which coherent X-rays are particularly sensitive. Several compounds stabilizing a CDW phase able to slide are presented, each with a different but clearly pronounced signature of the sliding state. Two main features emerge from this series of experiments which have been little treated until now, the influence of CDW pinning by the sample surfaces and the propagation of periodic phase defects such as charge solitons across the entire sample. Phase models describing the spatial and temporal properties of sliding CDWs are presented in the last part of this review.

Keywords: Charge Density Wave; Coherent Diffraction; sliding; charge propagation; incommensurate modulation; solitons

1. Introduction

As early as in the 1970s, several authors raised the importance of the phase in CDW systems [1,2]. Indeed, a Charge Density Wave (CDW) is described by a periodic modulation of charges $\rho(\vec{r}) = A\cos[2k_F\vec{r} + \phi(\vec{r})]$, where A is the amplitude and ϕ is the phase which denotes the position of the CDW relative to the atomic host lattice. As a matter of facts, external perturbations generally mainly affect the CDW phase. For instance, when submitting the system to an electric current, the threshold field above which the CDW depins from the atomic lattice and slides, leading to an additional current, is directly linked to local CDW phase variations, either through defects in the bulk [2], conversion processes at the electrodes [3] or pinning at the surface [4,5]. Although CDW deformation and phase shifts have been theoretically studied for a long time [6], the precise observation of the phase deformation has been missing until the advent of advanced X-ray diffraction techniques.

From a general point of view, the observation of all types of defects in condensed matter has always been challenging. Electron diffraction methods on thin samples or surface techniques such as STM are very efficient to observe crystal dislocations at the atomic scale. On the other hand, bulk experiments such as neutron or X-ray diffraction are also sensitive to defects but result from spatial averages which provide a global view of disorder at the macroscopic scale. Observing localized CDW phase shifts is a much more difficult task. In that case indeed, the phase shift does not concern the host lattice itself, but the periodic atomic displacement associated to the CDW, small in amplitude and which overlaps to the host atomic lattice. To some extent, this type of defect could be called a second order phase shift. The purpose of this review is to show how *coherent* X-ray diffraction can provide access to such peculiar phase singularities.

Experiments using coherent X-ray beams have been being developed continuously since the 90's thanks to the improvement of synchrotron sources [7]. Third generation sources are indeed able to deliver much brighter beams and smaller source sizes, allowing to take advantage of the coherence properties of the beam.

After presenting the methodological aspects from model examples, we will show how this technique has proved efficient to probe CDW phase shifts and their dynamics.

2. Coherent X-ray Diffraction to Track Phase Shifts in Condensed Matter

Speaking of *coherent diffraction* is actually a pleonasm. Indeed, the diffraction process originates from contructive interferences and is therefore a coherent phenomenon in essence. However, this expression is justified by the orders of magnitude involved. Indeed, the beam is defined by two characteristic lengths: the first one is related to the wavelength, and the second one to the relative angle of propagation. We thus define two coherence lengths, the longitudinal coherence length $\xi_l = \frac{\lambda^2}{2\Delta\lambda}$ and the transverse one $\xi_t = \frac{\lambda R}{2a}$, where λ is the beam wavelength, $\Delta\lambda$ the spectral width of the source, a is the numerical aperture of the source and R is the distance from the source. These two quantities have to be large enough to see interferences. But how large? It actually depends on the typical size of the object under consideration. For instance, in the case of classical X-ray diffraction on crystals, ξ_l and ξ_t have to be larger than the lattice parameters of the chosen crystal, which is always the case, even for laboratory X-ray sources. However, to obtain interference from larger objects, both coherence lengths must be scaled to the dimensions of the object. This is standard to get interferences from micron-size objects with visible light, using lasers typically, but harder to get with X-rays as ξ_l and ξ_t scale with λ. However, since the emergence of third-generation synchrotrons, micron-size values for ξ_t can be obtained thanks to micrometer source sizes and large distances between source and sample, while ξ_l in the micron range is achieved thanks to low bandwidth monochromators. Hence, we generally speak of *coherent diffraction*, when the coherence lengths of the X-ray beam are close to the size of the diffracted entities.

The diffraction pattern of a rectangular slit opened at few micrometers and leading to the expected cardinal sinus squared diffraction pattern (see Figure 1) is an illustration the phenomenon. The very good contrast of the interference fringes reveals the high degree of coherence obtained in the hard X-ray regime [8–10].

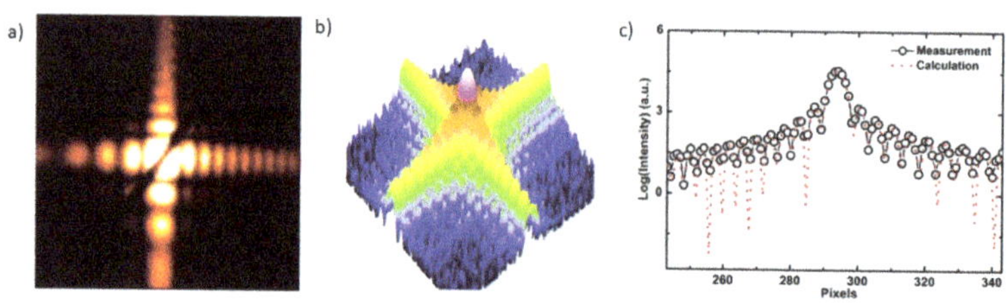

Figure 1. (**a**,**b**) CXRD pattern of a 7 keV X-ray beam by 5 μm square slits. (**c**) Profile of the diffraction pattern showing high-visibility fringes.

Many techniques emerged to take advantage of the coherent properties of beams produced by large-scale instruments, all based on the analysis of the interference patterns obtained by objects introducing a phase shift in the beam propagation. In condensed matter physics, any system that deviates from perfect crystallinity and/or is smaller than the beam size introduces such phase shifts and leads to interference patterns. Coherent X-ray diffraction (CXRD) thus opened the way to new opportunities, as the possibility to follow the fluctuation dynamics in condensed matter (called X-ray photon correlation spectroscopy) or to obtain the real-space image of the diffracted object using phase-retrieval methods (referred to as coherent diffraction imaging methods [11,12]).

There is however another possible application of the use of a coherent beam. Among all the possible defects encountered in condensed matter, some of them introduce phase shifts, as dislocations, for which coherent diffraction is particularly sensitive. Before discussing the case of dislocations in electronic crystals, let us illustrate the phenomenon with the textbook case of an isolated dislocation in a perfect crystal.

For example, an isolated dislocation loop can be stabilized in a silicon crystal after specific thermal treatments. Such a defect introduces phase-shifted domains on each side of the dislocation line. When a coherent X-ray beam probes regions containing such dislocation lines, interferences are observed (see Figure 2).

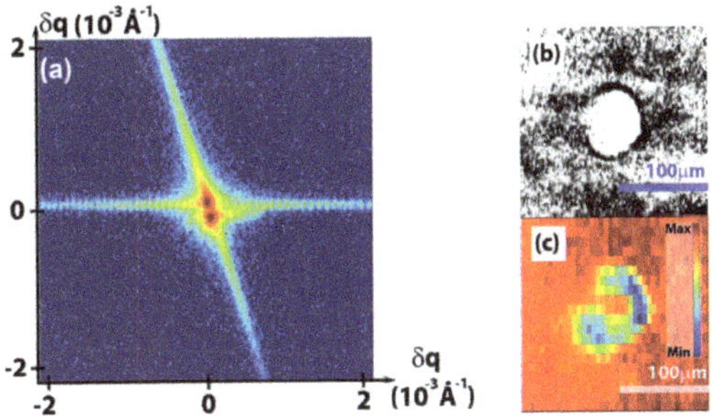

Figure 2. Coherent diffraction of a Silicon crystal displaying a dislocation loop (**a**) CXRD pattern obtained on the 220 reflection when the beam probes the dislocation line. (**b**) Topography image of a typical dislocation loop in this sample. (**c**) Image obtained in the same area as (**b**) by scanning the area with the 5μm coherent X-ray beam and plotting the intensity at the expected Bragg peak position. This experiment has been performed at the CRISTAL beamline of the SOLEIL synchrotron at E = 7 keV.

In perfects regions of the crystal, a well defined Bragg peak is observed. In contrast, when the beam probes the dislocation line, a destructive interference is observed, and the Bragg peak displays two side maxima (Figure 2a). The local minimum in between the two maxima can be tracked as a function of beam position on the sample to retrieve the full dislocation loop (Figure 2c). The resulting image is in agreement with images obtained by X-ray topography (Figure 2b). In addition, more details can be extracted of the CXRD pattern, especially from the oblique scattering line that reveals that these dislocation loops are dissociated into two partials (see [13] for more details).

2.1. Phase Shifts of Electronic Crystals Studied by Coherent X-ray Diffraction

The same methodology can be applied to electronic crystals (CDW and SDW) focusing the measurement on the satellite reflection associated to the periodic lattice distortion instead of the diffraction peaks associated to the host atomic lattice itself.

2.2. Isolated CDW and SDW Dislocations

First, electronic crystals can display their own phase defects such as dislocations that can be probed with a coherent beam, as described in the previous section. A first demonstration of a CDW dislocation was obtained in the blue bronze $K_{0.3}MoO_3$ [14]. This compound is a quasi-1D system made of chains of MoO_6 clusters along which an incommensurate CDW develops below 183K. The CDW being out-of-phase for adjacent planes, the CDW wavefronts are inclined with respect to the chain direction, as illustrated in Figure 3a.

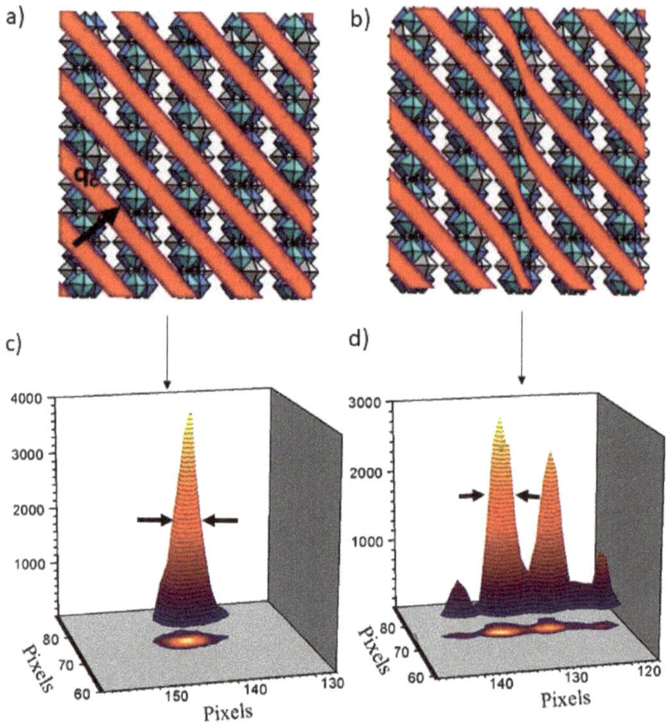

Figure 3. Schematic representation of the CDW wavefronts in the blue bronze (red lines) in the quasi-1D crystal structure (only MoO_6 octahedra are represented): (**a**) without a CDW dislocation and (**b**) with a CDW mixed-dislocation. Coherent diffraction pattern measured at two different sample positions at T = 75K. The single peak measured in (**c**) corresponds to a perfect CDW as represented in (**a**) whereas interference fringes in (**d**) are consistent with a CDW displaying a mixed dislocation, between an edge and a screw dislocation, as displayed in (**b**). Note that the widths of the two fringes are identical and equal to the width of the reflection associated to the perfect CDW.

In most regions of the sample, the satellite reflection associated to the CDW modulation displays a single peak (Figure 3c), indicating a long-range order greater than the beam size without CDW phase defect in the micron-sized probed volume. In other regions however, the CXRD pattern is split into two subpeaks with the same widths (Figure 3d). Similarly to the diffraction of a slit where all fringes have the same width (at half maximum intensity), all fringes display here the same widths. This is the typical signature of interference effects between two domains out of phase. This diffraction profile can be well reproduced by considering a mixed-dislocation of the CDW, between an edge and a screw dislocation, as schematically shown in Figure 3b).

Similar isolated phase defects can be detected in magnetic modulations, like SDW. This was observed in chromium, that exhibits both SDW and CDW modulations below T_N = 311K with associated satellites reflections at wave vectors $q_s = 2k_F$ and $q_c = 4k_F$ respectively. Using CXRD in a non-resonant magnetic mode, a characteristic splitting of the $2k_F$ satellite reflection associated to the SDW is observed at certain positions of the sample, while a single peak is visible at most other positions, and turned out to be in agreement with an edge-dislocation on the magnetic modulation (see Figure 4) [15].

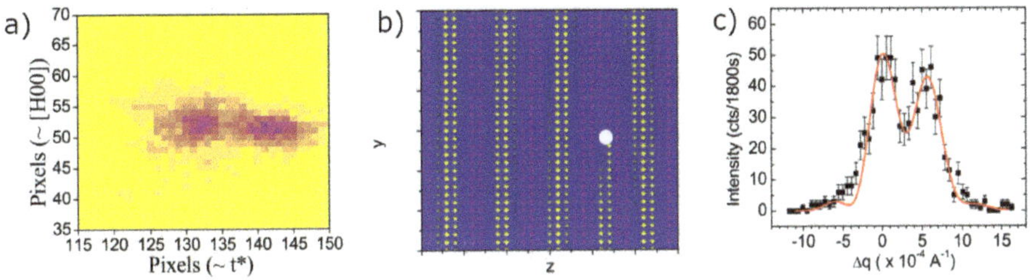

Figure 4. (**a**) CXRD pattern obtained on the SDW peak of chromium at a position where the peak is split, as seen on the CCD detector; (**b**) simulation of an edge dislocation line of the SDW. The white dot is the dislocation line position, and the yellow/red dots are a representation of the magnetic moments at each atomic position in the presence of a SDW dislocation; (**c**) red solid line: simulated CXRD pattern corresponding to the spatial arrangement shown in (**b**) represented on top of the experimental data (black squares).

2.3. *Coexisting SDW and CDW: Two Modulations with Highly Different Correlation Lengths*

Coherent diffraction experiments also allow to compare the state of disorder through speckles. Indeed, the number of speckles observed is, in first approximation, related to the number of defects. This property is particularly interesting in the case of chromium case that hosts two coexisting phases whose type of interaction has been much discussed.

If the SDW represents the main harmonics of the modulation with wavevector $\vec{Q}$, a CDW is concommitantly stabilized as its second harmonics at $2\vec{Q}$. However, while the magnetic instability is clearly due to the nesting of electron and hole pockets at the Fermi level with wave vector $\vec{Q}$, the origin of the CDW is far less understood. Several scenarii may account for its appearance. The first one relies on magnetostrictive coupling: the interaction between the SDW and the atomic lattice induces a CDW at $2\vec{Q}$ [16]. Another hypothesis involves a second nesting of unnested hole pockets following the SDW formation [17]. How to distinguish between this purely electronic or magneto-elastic scenarii? As coherent diffraction is very sensitive to local defects, similar profiles on the two reflections are expected in the case of the magnetostrictive origin of the CDW. However, comparing the singularities of the two phases is not an easy task in diffraction because the probed volumes are in general not equivalent. The way to avoid this is to use simultaneous diffraction geometry for the SDW and CDW reflections by placing simultaneously the $q_S = (0, 1-\delta, 0)$ SDW reflection and the $q_C = (1, 1-2\delta, 0)$ CDW reflection on the Ewald sphere (see Figure 5a). The images recorded at the maximum of the rocking curve for the CDW and SDW reflections using a 2D detector are displayed in Figure 5b,c respectively, with the same scale in reciprocal space.

The difference between the two diffraction patterns is striking: while the CDW reflection is broad and contains many speckles, the SDW reflection is as narrow as the direct beam. This reveals a high number of CDW defects while the SDW correlation length remains larger than the 10 μm × 10 μm probed volume. We can directly infer from these measurements that the the origin of the CDW is not directly linked to that of the SDW. The scenario relying on a purely magnetostrictive origin of the CDW does not hold while the results could be compatible with a band model based on a second order nesting [18].

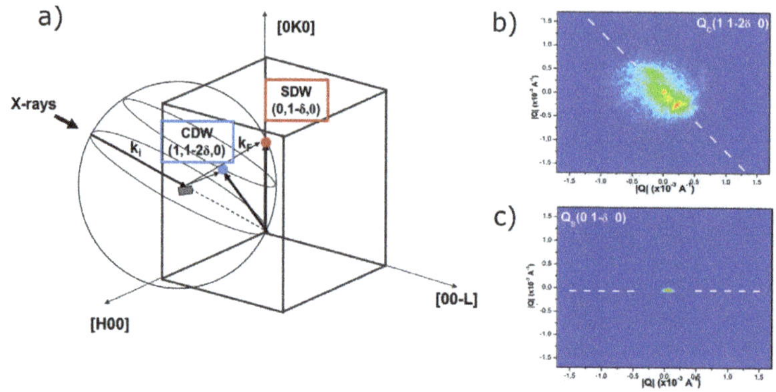

Figure 5. (**a**) Schematic drawing of the simultaneous diffraction experiment. Given an incident wave vector k_i, the sample is oriented so that both q_S and q_C satellite reflections are simultaneously located on the Ewald sphere, and therefore both fulfill the diffraction condition. Comparison of (**b**) CDW and (**c**) SDW satellite reflections in chromium by using coherent and simultaneous diffraction through the maximum intensity of the CDW (q_C=(1,1−2δ,0)) and SDW satellites (q_S=(0,1−δ,0)) (beam size = 10 µm × 10 µm). For most of the regions probed, the q_C satellite displays speckles, while no speckle is observed at q_S.

3. Dynamics of CDW Sliding Based on Phase Shifts Motion

The most spectacular property of incommensurate CDW systems is their ability to carry a collective current when submitting the sample to an external electric field (for a review, see [19,20]). Above a threshold bias current, an oscillating current is detected with a fundamental frequency as well as several harmonics [21]. Up to 23 harmonics have been observed in $NbSe_3$ [22]. This collective transport of charges through macroscopic sample has received a considerable interest for more than 35 years [19,23]. However, the understanding of the type of charge carriers involved in the phenomenon and their propagation mode still remains incomplete.

The first proposed scenario was based on the translational invariance of the incommensurate modulation allowing the whole CDW to slide over the atomic lattice without dispersion [24]. In fact, as we will see in the following, the sliding state is characterized by a strong distortion of the CDW. In addition, the CDW is an almost sinusoidal modulation as shown by diffraction experiments (the second harmonic is usually very weak in intensity) while transport measurements reveal a strong anharmonic signal [22] which suggests that the transport of charges in CDW systems is far more complex than a simple CDW translation. Another description considering the influence of defects, assumes a slowly varying phase $\phi(x)$ of the CDW interacting weakly with impurities [2]. The existence of the threshold field is thus well explained by considering an empirical bulk pinning potential, either strong or weak, depending on the type of defects and their concentration [23]. On the other hands, a strong electron-phonon coupling has been considered where the lattice itself plays the role of CDW pinning leading to discontinuities, in the phonon spectrum and atomic modulation [25]. A pure quantum tunneling through the sample was also proposed [26]. However, the most accepted theory, developed by Ong and Maki [3] and Gor'kov [27–29], deals with the CDW-metal junction at electrical contacts. The conversion of normal electrons from the metallic electrode into condensed charges in the CDW is made possible by climbing CDW dislocations at the interface. This so-called *phase slippage* and current conversion phenomena are in agreement with local resistivity measurements close to contacts [30]. In the phase slippage theories [3], impurities play a minor role, hidden in the tunneling coefficient. Note also that phase slippage mixed with quantum tunneling have been considered at low temperature [31].

The validation of either of these theories suffers, however, from the fact that it is very difficult to observe this phenomenon at the atomic scale. The aim of this review is to show

how the use of CXRD to observe CDW phase defects brings new insight on charge transport in CDW materials.

3.1. Dynamics of Sliding CDW Revealed by CXRD

The sliding state was historically observed by macroscopic resistivity measurements, but its signature in diffraction is also clear. Although each CDW system displays its own behavior, the sliding state is characterized in all cases by an increase of disorder below the threshold. Depending on the system under study, the type of disorder may take the form of creep, compression, expansion, rotation, or shear of the CDW wavefronts, with ordering processes by motion for larger currents or the appearance of an additional modulation appearing on top of the CDW.

As an illustration of the diversity of the phenomenon, let us first describe the behavior of the blue bronze $K_{0.3}MoO_3$ system under current (see Figure 6).

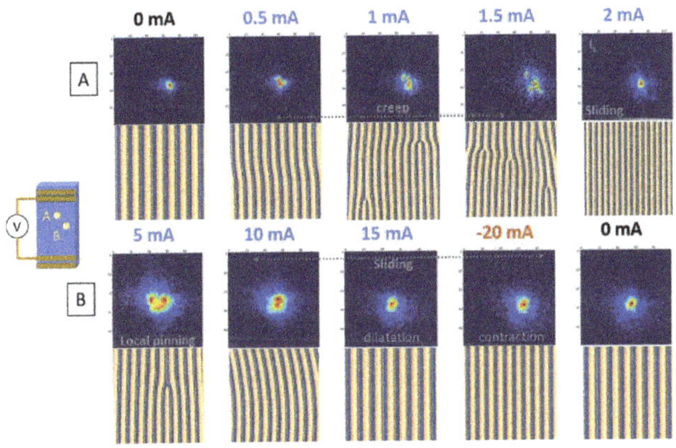

Figure 6. Coherent diffraction patterns of the $2k_F$ satellite reflection associated to the CDW of the blue bronze, at two positions A and B, while applying external dc currents. Each diffraction pattern corresponds to a sum of the full rocking curve and has been obtained for currents between 0 mA and 2 mA at position A and from 5 mA to 15 mA, then switched to −20 mA and back to 0 mA at position B. A sketch of the corresponding wavefront configuration in real space is displayed below each image, illustrating the creep regime below the threshold, the narrowing effect above the threshold due to the sliding motion, as well as dilatation and contraction of the CDW wavelength in the sliding state depending on the current direction. This experiment was performed at the CRISTAL beamline of SOLEIL synchrotron.

In this experiment, an external current was applied to the sample in a 4-probe configuration at 70 K, below and above the threshold current $I_S = 2$ mA. In most regions of the sample, the behaviour is similar to the one measured at position A in Figure 6: in the virgin state (I = 0 mA), the CDW reflection is made of a single peak, which accounts for a long-range order of the CDW. When current is applied, the CDW reflection broadens and displays speckles, which shows that the CDW loses its coherence even at very low currents, far below the threshold I_S, which is characteristic of the creep regime. Above I_S, when the macroscopic excess of current is observed by transport measurements, the CDW reflection gets narrower and accounts for a recovery of long-range order. However, this observation remains very local and is not homogeneous from one place to another. At another beam position (the beam size is few microns large) some intrinsic defects locally pin the CDW even above I_S, which gives rise to speckles (at 5 mA here, see Figure 6b). If current is further increased, the CDW can overcome the pinning center and recover its long-range order, here at I = 15 mA. By reversing the current, at I = −20 mA, this long-range order is maintained, and still apparent when the current is switched off [32,33].

Finer details revealing the effect of sliding can be detected. Probing the $2k_F$ CDW satellite with respect to external *dc* currents in the blue bronze reveals the existence of an extra modulation (see Figure 7).

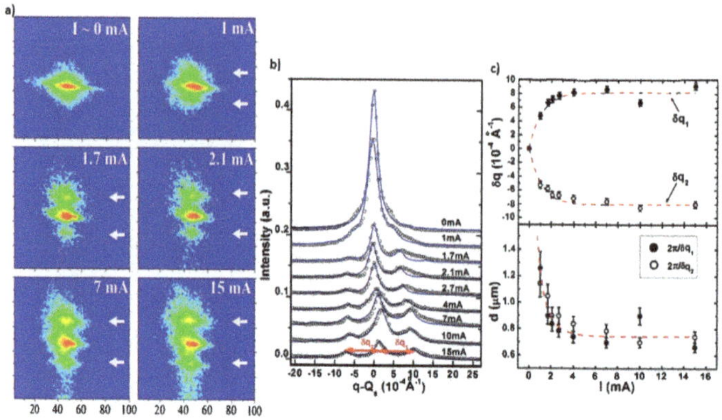

Figure 7. (**a**) 2D maps of the CDW satellite summed over the full rocking curve versus applied currents and (**b**) corresponding CDW profile along $\vec{b^*}$ (sum over the vertical CCD axis). (**c**) Evolution of the two secondary satellites position with respect to the $2k_F$ position (top panel) and the corresponding period in real space reaching more than one micrometer (lower panel).

In the sliding regime, the $2k_F$ satellite reflection displays secondary satellites along the chain axis which corresponds to the appearance of a new periodicity in the system, with periods in the micrometer scale i.e., 1500 times larger than the CDW wavelength that decrease with increasing currents (see Figure 7). We will come back to this experiment in the next chapter.

The sliding state is characterized by different features depending on the system under consideration. Figure 8 shows a comparison between two other sliding systems: the quasi two dimensional system $TbTe_3$ and the quasi one dimensional one $NbSe_3$, both probed by CXRD. Although the two systems do not display the same behavior, the diffraction patterns are very sensitive to the threshold current in both cases.

$TbTe_3$ samples are intrinsically much less ordered than the blue bronze or $NbSe_3$, leading to broad diffraction peaks and speckles. However, this disorder does not prevent the system from sliding [34] and the non-Ohmic conductivity is intimately linked to a strong distortion of the CDW.

The satellite reflection associated to the CDW shown in Figure 8a) displays speckles even without current. Below the threshold current I_S, the peak remains unchanged, but a visible shift in position is observed above I_S. This global shift corresponds to a rotation of the CDW wave vector in the sliding state (Figure 8a) [35]. Despite this reordering, one can still observe speckles surrounding the peak proving that the CDW remains in a disordered state above I_S.

The case of $NbSe_3$ is quite different. The diffraction pattern without current displays an almost single peak corresponding to CDW correlation lengths larger than the beam size, i.e., more than several micrometers in all directions. For small currents, far below the threshold, the satellite reflection displays an elongated shape along the transverse direction and is made of speckles (see Figure 8b).

Once the applied current exceeds the threshold current for sliding, speckles disappear in $NbSe_3$ leading to smooth diffraction profiles (see Figure 8b). This effect is more visible in Figure 9. The disappearance of speckles does not correspond to a decreasing number of CDW phase shifts but to time average, the counting time to get the image being longer than the characteristic time of CDW sliding. Indeed, the 10s acquisition time necessary to obtain one diffraction pattern is long compared to the phase shifts motion [36].

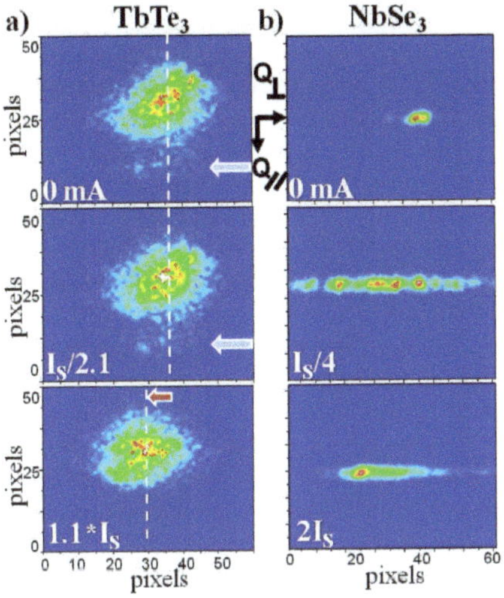

Figure 8. Coherent diffraction patterns of the $2k_F$ satellite reflection associated to the CDW versus external current, below and above the threshold current I_S, in (**a**) the quasi-two dimensional $TbTe_3$ system (for I = 0 mA, $I = I_S/2.1$ = 5 mA and $I = 1.1\times I_S$ = 12 mA) where the red arrow indicates the shift of the $2k_F$ reflection at I_S. (**b**) in the quasi-one dimensional $NbSe_3$ system displaying an elongated shape made of speckles below the threshold before refining above. The 2D images are a sum over several incidence angles through the maximum of intensity. Although the cutting plane is different in the two cases, the vertical direction of the camera is close to the $2k_F$ wave vector ($Q_{||}$) and the horizontal one is transverse to $2k_F$ ($Q_\perp$) in both cases [35].

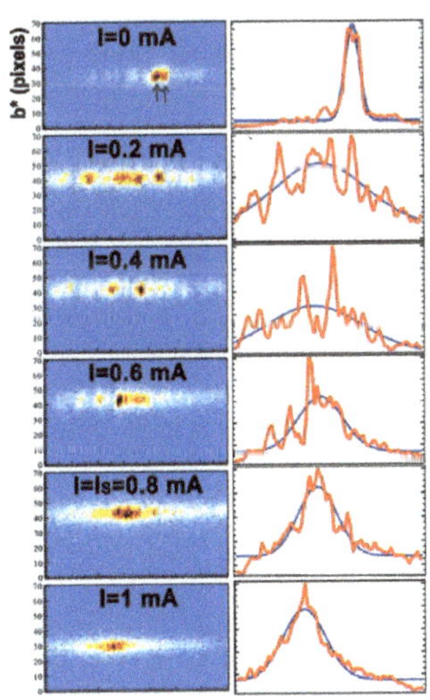

Figure 9. 2D diffraction patterns of the (0,1.241,0) satellite reflection under external current from 0.2 to 1.8 mA (left column), as well as the corresponding transverse profiles (right column) obtained after integration over the longitudinal direction. Speckles are observed even for weak currents and disappear above the threshold at I_s = 0.8 mA due to time average [36].

3.2. Microscale Shear Deformation of a CDW Induced by Surface Pinning

The limitation of the previous experiment is that the probe remains static and large, averaging the CDW over the whole illuminated region of the sample, and thus excluding the possibility to observe local variations of the CDW. The $NbSe_3$ system, however, is known to develop a continuous CDW deformation under current. Indeed, current-induced CDW deformations have been measured mainly close to the two electrodes by using a 50 μm × 50 μm X-ray beam along the entire sample length. The CDW appears to be compressed on one edge close to the electrical contact and expanded on the other, leading to a clear phase asymmetry in this direction [4,37]. These deformations were also observed by local resistivity measurements [38] and are consistent with those required for phase slip and CDW-to-normal carrier conversion at the contacts [4].

Although predicted in the 80's [6], at least in the close vicinity of the surface, the CDW shear deformation, which takes place along the direction perpendicular to current injection, had never been observed. This is explained by the fact that $NbSe_3$ samples have the shape of very thin wires, tens of micrometers wide requiring the use of a much smaller beam to observe transverse deformations without space averaging. In this regard, an X-ray micro-diffraction experiment has been performed in $NbSe_3$ versus applied *dc* currents. Four gold contacts were evaporated on a 39 μm × 3 μm × 2.25 mm single crystal glued on a sapphire substrate to perform four points resistivity measurements in-situ. Fast scanning diffraction technique allows us to map the CDW sliding across the $NbSe_3$ cross section. The precise $2k_F$ wave vector has been measured as a function of the X-ray beam position on the sample surface. As shown in Figure 10, 100 μm × 40 μm maps were probed with 1 μm resolution as a function of current. From these diffraction patterns, the CDW phase has been obtained by using a phase gradient method [39]. The maps in Figure 10 have been obtained using the gradient method by considering the map measured at I = 0.15 mA as the reference map. Indeed, due to hysteresis effects, it is always difficult to start the current injection from the true virgin state. However, we have considered that the reference map chosen was similar to the CDW initial state without current (see Figure 10 for more details).

In the upper part of the sample, in which current flows, a continuous deformation is observed from one lateral surface to the other, while the part in which no current flows remains unchanged. Like a guitar string plucked at both ends and subjected to a transverse force, the CDW bends in one direction or the other depending on the current direction. Despite the imperfections of the crystal, the CDW displays a continuous shear through the whole sample cross section, i.e., across 20 μm, which corresponds to more than 10,000 times its wavelength (λ_{CDW} = 14 Å). This continuous deformation spreading over such a large distance, and leaving both boundaries unchanged, emphasizes that a CDW is able to maintain its cohesion over macroscopic distances despite the local disorder. A CDW, a least in $NbSe_3$, is mainly pinned by the lateral surfaces and little by the bulk, in contradiction with bulk pinning theories [2].

Another indication that the surface has a dominant effect on CDW sliding is that the threshold current depends on the sample size and increases as it decreases. Resistivity measurements show that the threshold value diverges with decreasing sample length in $NbSe_3$ [40,41] and in TaS_3 [42]. Resistivity measurements have also shown that the threshold field is sensitive to the *lateral* dimensions of the sample, and increase with decreasing sample cross-section in $NbSe_3$ [41,43] but also in TaS_3 [44].

To describe this sample length-dependence of the threshold, a phenomenological relationship between E_{th} and sample length L_x can be established, involving CDW bulk impurity pinning [6]. Batistic et al. numerically found $E_{th} \approx 2.55 L_x^{-\alpha}$ where $\alpha = 1.23$ considering longitudinal pinning [40], but this, however, does not explain the constant E_{th} observe for large L_x. On the other hand, a more precise description of the compression-dilatation profile developing along the CDW direction in $NbSe_3$ has been obtained by considering nucleation processes of dislocation loops considering creep effect and an incomplete conversion process [45].

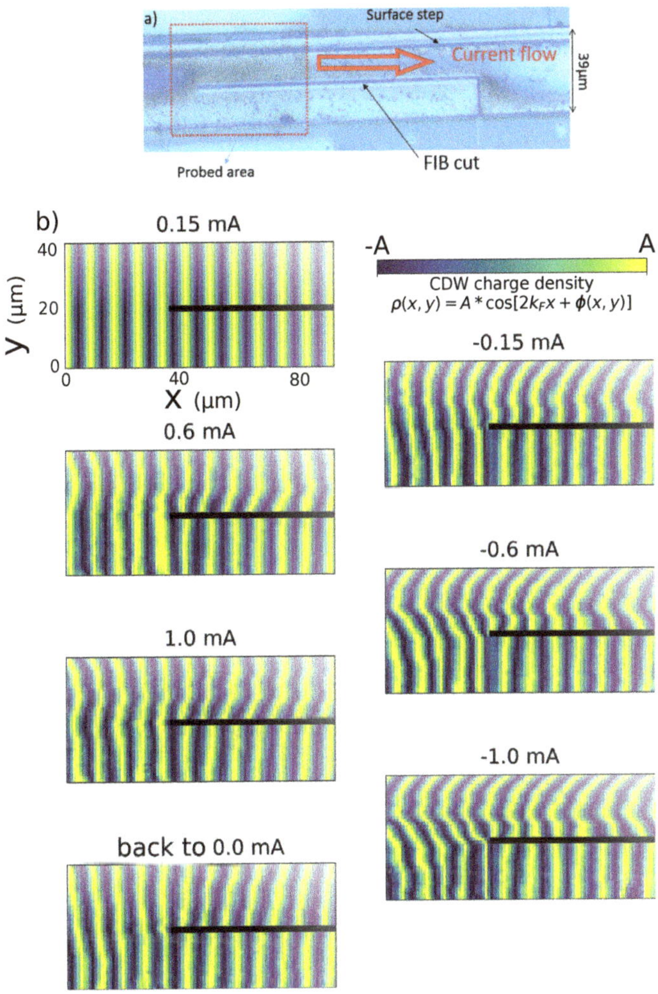

Figure 10. (**a**) Image of the $NbSe_3$ wire. The focused ion beam (FIB) cut forces the current to flow through the upper part of the sample only, and not in the lower part below the cut line. This geometry allows to simultaneously observe sliding and non-sliding areas from a single sample. The CDW wave fronts are displayed in yellow. The area mapped by the X-ray beam is indicated in red. (**b**) Evolution of the CDW in this region under increasing positive currents and decreasing currents down to negative values (I = {0.15, 0.6, 1, 0, −0.15, −0.6, −1}). The map at 0.15 mA has been considered as the reference map. For clarity, the period of the CDW (yellow wave fronts) has been considerably increased (in reality, the CDW period is λ = 14 Å) [39]).

This experiment obviously highlights the predominant role of pinning by lateral surfaces, until now neglected in previous theories (see Figure 10). In order to get the spatial dependence of the phase and the size-dependent threshold field $E_{th}(L_x, L_y, L_z)$, let us consider the known 3D CDW free energy [46,47]:

$$\mathcal{F}[\phi] \propto \int d^3\vec{r}\,\{c_x^2\phi_x^2 + c_y^2\phi_y^2 + c_z^2\phi_z^2 + \omega_0^2[1-\cos(\phi)] + \eta E x \phi_x\} \tag{1}$$

where c_x, c_y, c_z are the CDW longitudinal and transverse elastic coefficients, $\phi_j \equiv \frac{\partial\phi}{\partial j}$ are the phase derivatives, $V_{imp}(\phi) \equiv \omega_0^2[1-\cos(\phi)]$ is a standard emulation of the bulk impurity pinning potential [19,23] neglecting its randomness [2,23] and the last term corresponds to the CDW coupling to an applied electric field E with the longitudinal gradient ϕ_x, where η is a temperature-dependent coupling coefficient [47] and Ex is the applied electric potential. Guided by the experiment, we fix the phase at the electrical contact ($x = 0$ and $x = L_x$) and at the transverse surfaces. The corresponding boundaryconditions are :

$$\phi\left(\pm\frac{L_x}{2}, y, z\right) = \phi\left(x, \pm\frac{L_y}{2}, z\right) = \phi\left(x, y, \pm\frac{L_z}{2}\right) = 0, \quad (2)$$

The variational equation for the functional (1) (see Equation (4) below) was solved using the Green function and image charges method (see details in [48,49]). In the first order in β, the solution yields:

$$\phi(\vec{r}) \approx -\frac{32}{\pi^5} E\eta\beta \cos(\pi\frac{x}{L_x}) \cos(\pi\frac{y}{L_y}) \cos(\pi\frac{z}{L_z}) \quad (3)$$

where the coefficient β dependents on the sample size L_x, L_y, L_z as:

$$\beta = \frac{1}{\frac{c_x^2}{L_x^2} + \frac{c_y^2}{L_y^2} + \frac{c_z^2}{L_z^2} + \frac{\omega_0^2}{2\pi^2}}$$

As shown in Figure 11, Equation (3) for the phase satisfies the Dirichlet conditions (Equation (2)). The solution corresponds to a CDW shear in the central part of the sample and a compression or a dilatation of the CDW wave fronts at the two edges, in agreement with experiments.

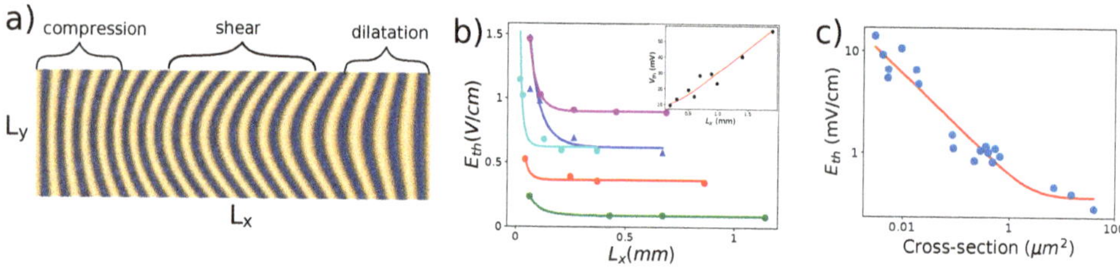

Figure 11. (**a**) CDW including the $\phi(x, y)$ obtained from Equation (3) displaying a shear effect with a curvature of the wave fronts in the middle part of the sample and a compression-dilation of the CDW period at the two electrical contacts. The CDW wavelength λ has been significantly increased for clarity (in reality $\lambda = 14$ Å in $NbSe_3$, that is $\lambda \approx 10^{-6} L_x$) (**b**) Threshold field E_{th} versus L_x and its corresponding fit using Equation (13) of reference [49]. The experimental dots were obtained in $NbSe_3$ (reproduced from [40]) and the blue triangles in TaS_3 (from [42]). (**c**) Threshold field E_{th} (blue dots) versus sample cross section in small o-TaS_3 samples (reproduced from [44]). The fit (red line) correctly reproduces the increase of E_{th} for decreasing cross sections and the asymptotic constant value for large cross sections [49].

The threshold dependence on L_x and L_y can be obtained by considering a threshold strain ϕ' leading to $E_{th} \propto 1/L_x$ and to a constant E_{th} at large L_x. Experimental data in Figure 11b) are shown with their corresponding fit using Equation (13) from Ref. [49]. The same equation was used to fit the evolution of E_{th} as a function of the sample cross-section $S = L_y L_z$ shown in Figure 11c). Furthermore, bulk impurity pinning was removed for those fits ($\omega_0 = 0$) showing that surface pinning alone is sufficient to explain the constant E_{th} at large L_x (see [49] for more details).

As a conclusion, without considering the empirical bulk pinning ($\omega_0 = 0$), and by only fixing the phase at a constant value on all surfaces, the global deformation of the CDW under current can be reproduced, including the dilatation and the compression close to electrical contacts [4,37], the wave front curvature in the middle part [39], and the threshold field dependence on sample length and cross section.

This observation raises questions about the very nature of this phase able to develop a continuous deformation across macroscopic distances in imperfect lattices containing many defects in volume [26]. We can also note that if the transverse deformation is clearly related to lateral surface pinning, the longitudinal one isn't since the CDW compression-dilatation at the electrodes is also due to the conversion of normal electrons from the metallic electrode into condensed charges [4,45].

3.3. Sliding CDW Based on a Traveling Soliton Lattice

In the previous chapters, we have shown several aspects of a CDW that are all related to sliding: CDW can stabilize dislocations at rest (see Figure 3) [14], those phase shifts are mobile above the threshold, leading to the disappearance of speckles in blue bronze (see Figure 6) and in $NbSe_3$ (see Figure 9) [36]; the CDW displays a strong distortion (longitudinal and transverse) in $NbSe_3$ and an additional periodicity appears on top of the CDW when sliding in $K_{0.3}MoO_3$ blue bronze (see Figure 12) [32]. The excess of current observed in CDW systems above the threshold could be related to all these observations.

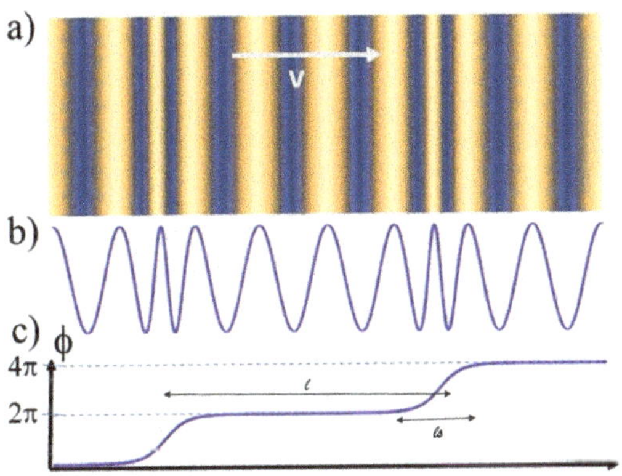

Figure 12. Sketch of the soliton lattice in real space. (**a**) CDW in the presence of a soliton lattice with period l corresponding to periodic 2π phase shifts spread over a distance l_s and propagating along the applied field (**b**) the corresponding electronic density and (**c**) the phase ϕ profile [50].

Let us come back to the extra modulation observed in the sliding regime of the blue bronze (see Figure 7). This modulation can be understood as the presence of a soliton lattice in translation. For this, let us first consider a crude model considering the influence of defects through an interaction which couples the pinning potential and the phase ϕ [2]. The free energy leads to the following equation of motion in 1D [23]:

$$\frac{\partial^2 \phi}{\partial t^2} - c_x^2 \frac{\partial^2 \phi}{\partial x^2} + \eta \frac{\partial \phi}{\partial t} + \omega_0^2 \sin(\phi) = F \tag{4}$$

where $F = \frac{2c_\phi^2 e E}{\hbar v_F}$ is proportional to the applied electric field, $c_x = \sqrt{m/m^*}\, v_F$ is the phason's velocity and ω_0 the pinning frequency. We also add an effective damping term $\eta \frac{\partial \phi}{\partial t}$ to mainly take into account the coupling between CDW and phonons. The $\sin(\phi)$ term is not linearized here allowing for abrupt phase variations. The usual non-perturbed sine-Gordon equation (for which $F = \eta = 0$) is known to admit soliton solutions. However, soliton excitations are quite robust and survive the inclusion of a reasonable external force and dissipation keeping their topological properties although the soliton shape is slightly modified [51]. Let us now solve Equation (4) considering that the phase $\phi(x,t)$ contains two terms: a slowly varying phase $\phi_0(x)$ and a dynamical part $\phi_1(x,t)$ where ϕ_1 varies much more rapidly than the static one. The static part $\phi_0(x)$ can then be calculated by averaging Equation (4) in time:

$$\left\langle \frac{\partial^2 \phi_0(x)}{\partial x^2} \right\rangle_t = (\frac{\eta\,\pi}{e} j - F)/c_\phi^2, \tag{5}$$

leading to a quadratic variation of the phase $\phi_0(x)$ ($j = e/\pi \langle \partial\phi_1/\partial t \rangle$). The CDW, $\rho = \rho_0 \cos(2k_F x + \phi(x,t))$, pinned at both ends at the metal/CDW junction, is compressed at one electrode and stretched at the other in agreement with experiments [4,37,45]. The excess of current in the sliding regime $j = \frac{e}{\pi}\frac{\partial \phi}{\partial t}$ is constant far from electrodes as observed by numerous transport measurements [38,52–55].

The dynamical part $\phi_1(x,t)$ obeys the sine-Gordon equation and is submitted to an effective force including friction. Considering the periodic nucleation of CDW dislocations at the electrode [31], we obtain a train of solitons [51]:

$$\phi_1(x,t) = \delta + \sum_{n=-\infty}^{\infty} 4 \arctan\left(\exp\left(\frac{x - v_S t - ln}{l_S\, \gamma(v)} \right) \right), \tag{6}$$

where l is the distance between successive solitons and $l_S = c_x/\omega_0$ their extension (see Figure 13). Overlapping effects between solitons are neglected ($l/l_S > 2$) [50]. Note also that the soliton lattice reaches a stationary sliding velocity v_s proportional to the electric field E. Another expression for this soliton lattice, using the elliptic Jacobi function and giving essentially similar results, will be discussed later (see Equation (14)).

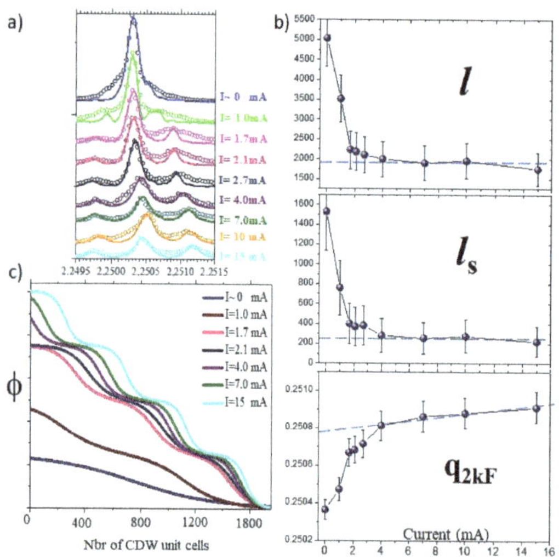

Figure 13. (**a**) Fit of the experimental data from the soliton lattice model (Equation (6)). (**b**) Fitted parameters l_s and l versus currents and (**c**) profile of the corresponding phase ϕ showing an increase of the soliton density for increasing currents [50].

This scenario of a soliton lattice traveling through macroscopic samples explains the main features of the collective transport of charges in CDW systems. In this approach, once created, these topological objects propagate without dispersion and with a remarkably long lifetime. Furthermore in this theory, the propagation velocity increases while the lattice period decreases with increasing applied fields and these 2π solitons carry a localized charge. All these properties are in agreement with observations. It explains the well-defined frequency of the additional pulsed current despite the macroscopic dimensions of samples. Indeed, the charges remain spatially correlated despite the distance and the presence of disorder thanks to the robustness of these topological objects. The existence of the soliton lattice is also in agreement with the appearance of an additional periodicity observed by X-ray diffractioni experiments in the sliding regime of the blue bronze (see Figure 7). It also explains why the fundamental frequency of the additional current increases with increasing currents (since $v_s \propto$ E) and why the additional periodicity observed in the blue bronze decreases with increasing currents.

4. Effects of Solitons Observed by Diffraction

In this chapter, some theoretical aspects of the physics of solitons will be described in relation to X-ray diffraction experiments. Only the unidirectional case, along the $2k_F$ direction, is considered.

CDWs are subject to stresses that can come from a variety of sources, such as surface effects inducing pinning and shear [39] and/or changes in transition temperatures [56]. CDWs can also be stressed by surface doping, surface steps [57], proximity to commensurability, various structural defects like twins, domain walls [5], a constraint geometry [58] or an imbalance of normal and collective currents near junctions in the sliding regime [45]. The stress can easily exceed an elastic limit leading to the appearance of topological defects. For commensurate or near commensurate CDWs, the associated stresses can leave particular fingerprints such as a soliton lattice or a system of random solitons [32,59–61]. All these effects can, in principle, be observed by local probes such as STM or by space-resolved X-ray beam as previously described in this article (see also [5,38,45,57]) although the interpretation is not always obvious.

The intensity of the elastic scattering of the CDW with a distorted phase $\phi(\mathbf{r})$ is given by the expression:

$$I(\mathbf{q}) = I_0 \left| \int S(\mathbf{r}) \cdot e^{i\mathbf{q}\cdot\mathbf{r}} \cdot e^{i(\phi(\mathbf{r})-\phi(\mathbf{0}))} \cdot d\mathbf{r} \right|^2 \tag{7}$$

where I_0 is the normalization constant, and $S(\mathbf{r})$ is the correlation function describing an intrinsic disorder or inhomogeneity. X-ray experiments usually recover the "square Lorentz" intensity profiles [62] which correspond to a simple exponential decay in real space: $S(\mathbf{r}) \sim \exp(-|\mathbf{r}|/\eta)$ where η is the correlation length.

The M-fold commensurate CDW interacts with the underlying lattice via the commensurability energy which, for a small CDW amplitude A, can be written as (see e.g., [23]):

$$W_{com} = \alpha(1 - \cos M\phi)\ ,\ \alpha = A^M \tag{8}$$

Cases $M = 2, 3, 4, 8$ are known in various CDW materials with $M = 4$ for $NbSe_3$ and $M = 8$ for Blue Bronzes; the case $M = 1$ is to emulate the interaction with host impurities if we ignore the randomness of their positions. With primary contributions to the CDW energy $\propto A^2$, the commensurability effect is weak for $M > 2$, affecting only the low energy deformations of the phase, leaving the amplitude A almost unchanged.

For an exactly commensurate system, W_{com} results in locking of the CDW phase ϕ at values that are multiples of $2\pi/M$. The actual band filling and/or local perturbations of the concentration of condensed electrons enforce small deviations q from commensurability. The overall phase displays increments $\Delta\phi = qL$ over a length L and the phase jumps $N = \Delta\phi M/2\pi$ times along the sequence of commensurability plateaux, forming a lattice of solitons (see experimental Figure 13, and theoretical Figures 14). As previously described in the experimental section, the soliton lattice is characterized by two lengthscales, the distance between solitons $l = \frac{L}{N} = 2\pi/(Mq)$ and the soliton width ξ (similar to l_s in the previous arctangent model, see Equation (6)). The period l depends on band filling which makes it temperature-dependent due to thermal activation of carriers [63,64]. The single soliton width $\xi \propto (C_x/\alpha)^{-1/2}$ is jointly determined by the energy α and the CDW rigidity parameter C_x. It also depends on the carriers concentration due to the so-called effect of Coulomb hardening [19,65–67]. The basic model describing effects of the commensurability upon both the ground state and the spatio-temporal evolution of the CDW phase relies upon equation [51]

$$m^*\frac{\partial^2\phi}{\partial t^2}+\gamma\frac{\partial\phi}{\partial t}-C_x\frac{\partial^2\phi}{\partial x^2}+\frac{e}{\pi}E^*+F_{com}=0\,,\ F_{com}(\phi)=\alpha M\sin(M\phi) \tag{9}$$

whose parameters will be discussed later. The simplicity of the model and availability of exact solutions (the pure sine-Gordon limit with $\gamma = 0, E^* = 0$) provoked a great amount of studies through decades ([50] and references therein). However, the precise measurement of the different parameters of this equation is far from being simple. The effective *mass density* m^* is accessible by optical measurements though the parameter α which is related to the measurable pinning frequency ω_{pin} as $\alpha M/m^* = \omega_{pin}^2$. With regards to the damping coefficient γ, it is related to the measurable conductivity of the collective CDW sliding σ_c as $\gamma = e^2/(s\sigma_c)$ where s is the area per chain. The definitions of σ_c, the stiffness C_x, and the effective driving field E^* require a comprehensive approach to CDW physics, including such essential elements as Coulomb forces and normal carriers. The full microscopical approach with some applications was reviewed in [19], the transparent equations have been derived in [68] and the reasonably simplified version can be written as:

$$\frac{1}{\pi}\frac{1}{s\sigma_c}\partial_t\phi-\frac{1}{\pi}\rho_c/\rho_n\partial_x^2\phi+\frac{F_{com}}{e^2N_F}=-\frac{J(t)}{s\sigma_n}\,,\ \frac{1}{\sigma_c}=\frac{1}{\sigma_{CDW}}+\frac{1}{\sigma_n} \tag{10}$$

which is comparable, term by term, to Equation (9) allowing to interpret its components. The microscopically-derived temperature-dependent parameters are the normal and condensed densities ρ_n and $\rho_c = 1-\rho_n$, with $\rho_c(T_c) = 0$. They are normalized to $\rho_c(0) = 1$, and hence $\rho_n(0) = 0$. The second term on the left-hand-side describes the enhancement of the phase elasticity C_x and the ratio ρ_c/ρ_n controls the rigidity vanishing, while approaching the metallic state at $T \to T_c$. It also explains the Coulomb hardening between condensed charges which dramatically increases with freezing out of screening by normal carriers when $\rho_n \to 0$ at $T \to 0$. The right-hand-side of Equation (10) gives the driving force E^* as the total monitored current $J(t)$ divided by the normal conductivity σ_n alone. The total current is additive: $J = J_n + J_{CDW}$ with the intrinsic CDW current (per chain) being $J^1_{CDW} = -(e/\pi)\partial_t\phi$.

These complications, not quite intuitive, arise from the complex interplay of Coulomb interactions with screening facility of normal carriers. The effects are particularly strong for solitons which are actually the walls crossing the sample; their charge density generates a constant electric field which must be screened by the cloud of normal electrons whose width is the screening length $\propto 1/\rho_n$. This cloud deforms or moves together with the phase evolution thus contributing to both elastic energy and friction.

The boundary conditions applied to the solution of Equation (9) also require special attention. It is already known that without pinning, the solution in the dissipative limit is: $\phi = (px^2 - rt)$ where p and r are related as follows: $C_xp + \gamma r = eE^*/\pi$. In the opposite case, without boundary conditions, there is a freedom to redistribute the action of the driving force among the elastic $\sim p$ and viscous $\sim$ r contributions. We will demonstrate below the difference in behavior under different boundary conditions, both physically motivated.

Experiments performed on CDW systems show that the sliding motion is essentially dissipative. According to optical data (see e.g., [23]), the CDW response is overdamped for frequencies below 10 GHz, while the characteristic frequencies of CDW sliding, as measured by Narrow Band Noise (NBN) measurements, do not exceed 100 MHz. We must therefore keep only the derivatives in the dynamical equation (Equation (9)). The resulting equations can only be studied numerically, and is presented in the following.

The important question that arises is why the damping coefficient γ is so large, as shwon by the experiments dealing with the interaction of CDW with phonons. The answer also deserves to be integrated in the context of the physics of solitons, a now local phenomenon, caused by microscopic defects. Indeed, their presence is known to destroy the long-range CDW order. However, more importantly for us here, they provide an evolving metastable order, whose relaxation dissipates the sliding energy.

For the so-called local or strong pinning, the answer is explicitly known (see [69]). For small velocities, it looks like

$$\gamma = 4n_i F_\pi \tau_\pi \ , \ F_\pi = \frac{1}{2}\frac{d\Delta E}{d\theta}\bigg|_\pi \tag{11}$$

where n_i is the linear concentration of impurities, τ_π is the relaxation time of the local metastable state, F_π is the phase restoring force produced by the impurity, ΔE is the energy term dependent on the phase mismatch θ, and τ_π is the relaxation rate of the metastable state. More details are given in the last section.

In the following, we will first consider randomly-distributed static solitons, then a regular lattice of solitons, either static or moving under the effect of an external electric field, and finally the solitons in space and time domains appearing in the course of phase slips.

4.1. Rare Random Solitons

Consider a CDW system containing few solitons, with a mean concentration $c = 1/l$, trapped by impurities or thermally-scattered at $T > T_c$, so that their positions x_i are random. Each soliton produces a phase shift $\chi = 2\pi/M$. We shall keep χ to be arbitrary allowing also to include the effect of charged impurities, providing Friedel phase shifts even in absence of solitons (see [62] and refs. therein). The correlation function in (7) can be calculated via the Poisson distribution for a number of defects n found within the interval $(0, x)$. We have

$$D(x) = \left\langle e^{i[\phi(x)-\phi(0)]} \right\rangle = \sum_n e^{i\chi\cdot n\cdot sgn(x)} \frac{(c|x|)^n}{n!} e^{-c|x|} = e^{-|x|c(1-\cos\chi)} e^{ixc\sin\chi}. \tag{12}$$

In q space, we find a single symmetrical Lorentzian peak with the width $c(1-\cos\chi)$ and located at the displaced position $\delta q{-}c\sin\chi$:

$$I(q) = R\frac{2c(1-\cos\chi)}{(q-c\sin\chi)^2 + 4c^2(1-\cos\chi)^2} \ ; \chi = \frac{2\pi}{M} \tag{13}$$

where R is the material parameter. If $\chi \ll 1$, the shift is just given by the mean stretching $\delta q \approx c\chi$. Even small, the shift is nevertheless visible because the broadening $c\chi^2$ is even smaller. On the contrary, when the shift is maximal ($\delta q = c$ at $\chi = \pi/2$ for $M = 4$), for the unitary limit of the impurity potential, the peak position is no longer well-defined since its broadening is of the same order of magnitude as the shift.

4.2. Static Lattice of Solitons

Without driving force, $E^* = 0$, the static soliton of Equation (9) yields the regular soliton lattice whose form is known analytically:

$$\phi(x,k) = \frac{2}{M} am\left[\frac{M}{2k}\frac{x}{\xi}, k\right] \ ; \ l = \frac{4kK(k)\xi}{M} \ ; \ \xi = \left(\frac{C_x}{\alpha}\right)^{1/2} \tag{14}$$

where l is the distance between solitons, ξ is the characteristic soliton size, $am[\tau, k]$ is the elliptic Jacobi function, and $K(k)$ the complete elliptic integral of the first kind. This expression is more precise than the one used previously (see Equation (6)), although it may be less intuitive.

The intensity $I(q)$ is determined by Equation (7) containing the phase ϕ given by Equation (14). The results are shown in Figure 14. When the soliton size is large, for $\xi \sim l$ ($k \ll 1$), the phase ϕ grows almost linearly. In that case, the intensity $I(q)$ is simply shifted in q-space from the commensurability point Q_0 to $Q_0 + 2\pi/l$. In the opposite strongly non-linear case, where ϕ contains abrupt phase shifts ($\xi \ll l$), the solitonic superstructure leads to the formation of non-symmetric peaks, spaced by a wave vector depending of $1/l$ (Figure 14).

4.3. Lattice of Solitons Submitted to an External Electric Field

Consider now the static and the stationary states of the solitonic lattice under the applied homogeneous field E^*. Recall that the CDW sliding is overdamped, and so we consider the behavior of the phase ϕ described by Equation (9) in the limit $m^* = 0$. We numerically solve the differential equation by considering two different physical situations:

(i) The CDW condensate is considered as isolated, with a conserved total charge and submitted to boundary conditions fixing the phase increment:

$$\Delta\phi = \phi(L,t) - \phi(0,t) = 2\pi N_s/M \tag{15}$$

where N_s is the number of solitons over the chain length L present before the field is applied.

(ii) The CDW condensate without boundary conditions, only the initial phase distribution $\phi(x,0)$ is defined.

Independently of the boundary conditions, the CDW state is not static anymore for too large E, exceeding a critical electric field E_c. In the specific case of the rigid approximation, where $\partial_x\phi = 0$, the potential energy $-\alpha\cos(M\phi) - eEx/\pi$ behind the Equation (9) looses its minima above the critical electric field $E_{c0} = M\alpha$. In the more general case, where $\partial_x\phi \neq 0$, the threshold field E_c is different and our numerical solution shows that the actual threshold field is lower than the rigid case, with $E_c \approx 0.7E_{c0}$, being reduced by the allowed elasticity $\partial_x\phi \neq 0$.

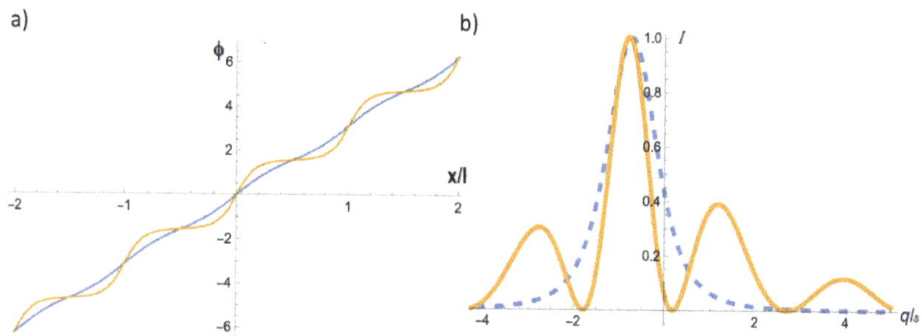

Figure 14. (**a**) The phase ϕ grows almost linearly for $\xi = 0.44l$ (blue line) or displays abrupt phase shifts $\xi = 0.04l$ (yellow line) (**b**) Corresponding normalized diffraction profile of the sattelite reflection associated to the CDW displaying the two types of solitonic lattice for $\xi = 0.44l$ (dashed blue line) and for $\xi = 0.04l$ (yellow line). The shift of the maximum of intensity is related to the average slope of the phase, which is equal in both cases here.

For the case (i) of the closed system, the space-time distribution of the phase $\phi(x,t)$ is presented in Figure 15 for low and high electric fields.

We see that the static solution corresponding to the lattice of somehow deformed solitons is reached at sufficiently long times. At the critical field, the soliton lattice starts to move and no static limit can be reached. Figure 16a shows the space dependence at a given point t_0 of $\phi(x,t_0)$ for various electric fields. In the sliding regime, the amplitude of undulations diminishes. At larger E, the profile of $\phi(x,t_0)$ approaches a plateau while the solitons are expelled in favor of a steep rise of ϕ near a boundary. The scattering intensity $I(q)$ is presented in Figure 16b for a subcritical electric field.

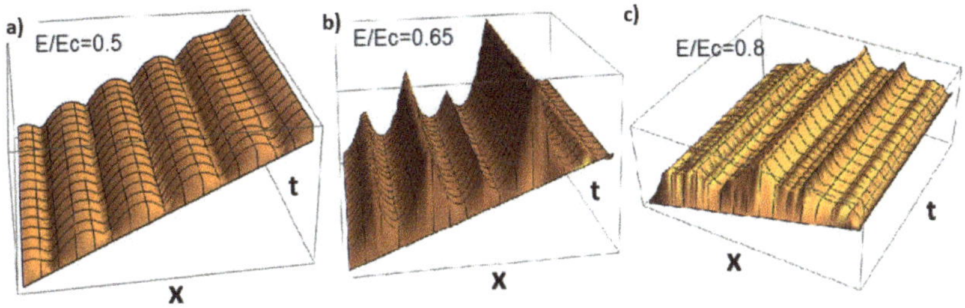

Figure 15. Space-time distribution of the phase ϕ for various electric fields: (**a**) $E = 0.5E_c$, (**b**) $E = 0.65E_c$ and (**c**) $E = 0.8E_c$.

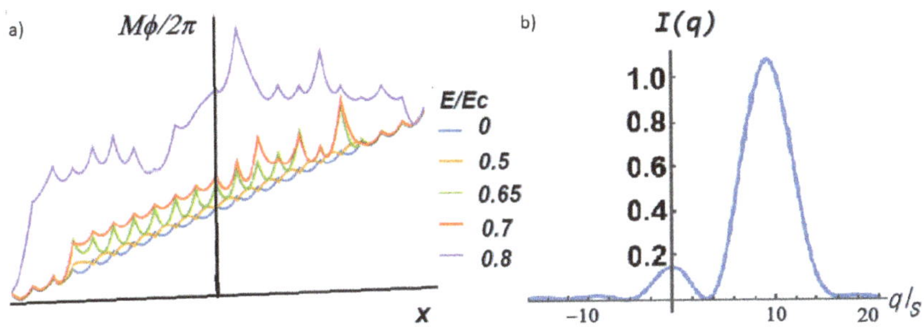

Figure 16. (**a**) Space distribution of the phase $\phi(x)$ at a fixed time for various electric fields. (**b**) Scattering intensity profile for an intermediate field $E/E_c = 0.5$.

For the case (ii) of the open system, we imply the initial conditions corresponding to the initial solitonic distribution:

$$\phi(x,0) = am(x,k) \tag{16}$$

The results of the numerical solution are presented in Figures 17 and 18. At short times, the soliton lattice profile $\phi(x,t)$ changes and becomes non-symmetrical. The soliton width increases with increasing applied fields, and the solitons start moving. The density plot of $\phi(x,t)$ is presented in Figure 17.

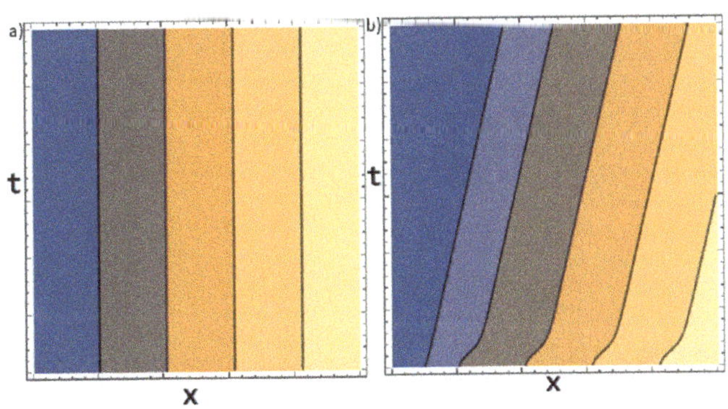

Figure 17. Contour plots of phase ϕ for (**a**) $E = 0$ and (**b**) $E = 0.15E_c$.

At zero electric field (Figure 17a), the initial periodic profile $\phi(x,0)$ is kept statically, but it starts to move under applied external electric fields with no threshold. The contour lines $\phi(x,t)$ bend and the distance between solitons changes not only in space but also in time (Figure 17b). The behavior of the phase in the middle part of the sample, $\phi(L/2,t)$, for various electric fields is presented in Figure 18a; the intensity $I(q)$ at an intermediate electric field is presented in Figure 18b.

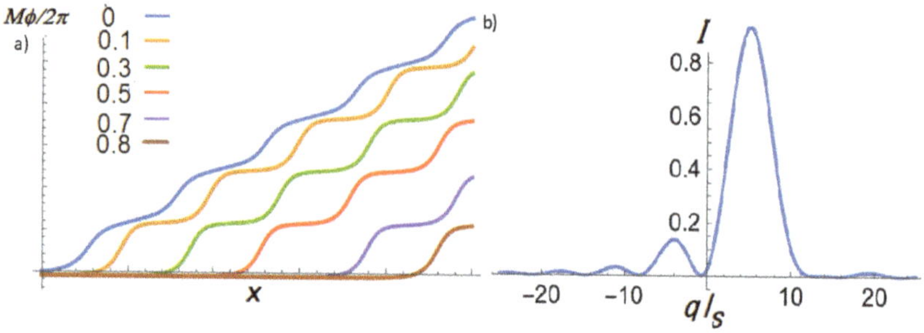

Figure 18. (**a**) Space distribution of the phase $\phi(x)$ at a fixed time for various electric fields. (**b**) Scattering intensity profile for intermediate field.

In this complex CDW sliding process, many questions can be answered by considering the phase-only framework, like how solitons enter or leave the sample, how they are (re)created near junctions, how the total soliton number N_s changes with temperature for example, or how the equilibrium wave vector evolves (see [70] and the first part of this review concerning diffraction experiments). However, a change of N_s, or more generally of the total phase increment $\Delta\phi$, is required for so-called phase-slip processes, which correspond to a kind of space-time vortices. As a consequence, it is required for the amplitude A of the CDW order parameter $\Psi = A\exp(i\phi)$ to vanish in the vortex core [49]. Thus, the equations must be generalized and include the function $A(x,t)$. The commensurability energy has to be generalized as $W_{com} \propto -\alpha(\Psi^M + \Psi^{*M})$ and the following additional equation for A has to be considered:

$$\kappa_x \partial_x^2 A + \kappa_x A(\partial_x \phi)^2 - A + A^3 - \alpha M\left(\Psi^{M-1} + \Psi^{*M-1}\right) = \tau^{-1}\partial_t A \tag{17}$$

where κ_x and τ (the amplitude relaxation rate) are some constants. In Equation (9), we must take into account that $\gamma \propto C_x \propto A^2$ and $\alpha \propto A^M$. The vorticity can be obtained only in invariant variables, so the phase derivative in Equations (9) and (17) must be generalized as $\partial_t\phi \to \omega = Im(\Psi^*\partial_t\Psi)/A^2$, $\partial_x\Psi \to q = Im(\Psi^*\partial_x\Psi)/A^2$ with the phase being restored as $\phi = \int^t \omega(t')dt'$. The numerical solution was performed in terms of components u, v of $\Psi = u + iv$. The boundary conditions are given in terms of q ; in view of the local electroneutrality condition $(q/\pi) + n = 0$, q specifies the concentration of normal electrons n, and thus their chemical potential which is the standard assumption.

Figure 19 shows an example of numerical solutions for $M = 1$. We see a stratification among the pinned bulk where the phase is nearly constant and the sliding stripes near junctions where the phase evolves by 2π pulses (solitons in the time domain). The regions are separated by a periodic array of vortices in time, and form as a wall in the x direction, which can be viewed as space-domain solitons. The plot of the amplitude shows the sequence of nodes: as expected, $A(x,t)$ goes to zero at the space-time vortex centers.

In conclusion, the commensurability solitons can be observable in CDWs with a sufficient concentration of normal electrons. Particular manifestations near junctions are challenging for space-resolved studies, particularly coherent micro-diffraction. These intriguing spatial and temporal effects require the use of both space and time-resolved techniques to be observed.

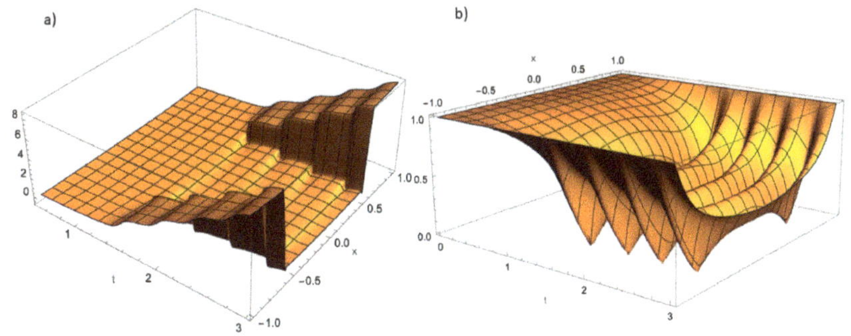

Figure 19. Phase slip processes. (**a**) Space-time distribution of the phase $\phi(x)$ and (**b**) of the amplitude $A(x,t)$.

4.4. Generation of Pairs of Solitons by an Impurity and Resulting CDW Viscosity

Consider a system of interacting CDW chains with a point impurity located at position $\vec{r}_i$, $x = x_i$ on the chain $n = 0$. We can write the Hamiltonian as

$$H = \int dx \left\{ \sum_n \left[\frac{1}{2} C_{\parallel} (\partial_x \phi_n)^2 - \sum_m C_{\perp} \cos(\phi_n - \phi_m) \right] - V \cos(\phi_0 + \vec{Q}\vec{r}_i) \delta(x - x_i) \right\} \quad (18)$$

where $C_{\perp}$ is the interchain coupling and V is the impurity strength. The 2π periodicity of the pinning energy allows to skip the 2π quanta in ϕ_0 to optimize the total energy. Moreover, the 2π periodicity of the regular energy in Equation (18) allows for interchain $\pm 2\pi$ solitons. For the soliton centered at position X, the phase profile $\phi_s(x - X)$ describes stretching/dilatation by one period along the defected chain relative to the surrounding ones. The soliton is distributed over the length $\xi \sim \sqrt{C_{\parallel}/C_{\perp}}$ and costs the energy $E_s \sim \sqrt{C_{\parallel} C_{\perp}}$, the two terms defining the equilibrium concentration of solitons $n_s \sim \exp(-E_s/T)$.

The energy should be minimized over $\phi(x)$ with the asymptotic condition $\phi \to \bar{\phi}$ at $|x - x_i| \to \infty$ where the mean phase in the bulk $\bar{\phi}$ can be time-dependent. It is convenient to keep the local value $\phi_0(x_i)$ fixed and optimize it only at the end of the calculation. Then the pinning center can be described by a single degree of freedom ψ_i and monitored by another single one θ_i. Let us define the local mismatches of phases relative to the bulk value $\bar{\phi}$:

$$\psi_i = \phi(x_i) - \bar{\phi}\,,\ \theta_i = -Qx_i - \bar{\phi}\,, \partial_t \bar{\phi} = \omega \quad (19)$$

Henceforth, the index i will be omitted.

Quantitative results can be obtained within a short-range model (Equation (18)). If we consider that only the central chain $n = 0$ (passing through the impurity) is perturbed while its neighbors stay at $\phi_{n \neq 0} = \bar{\phi}$ homogeneously, then the energy functional can be simplified as:

$$\int dx \left[\frac{1}{2} C_{\parallel} (\phi')^2 - C_{\perp} \cos(\phi) - V \cos(\phi - \theta) \delta(x) \right] \quad (20)$$

Its extremum is the function $\phi(x) = \phi_s(x - X) - \phi_s(x + X)$ where $\phi_s(x)$ is the standard sine-Gordon soliton shape and X is fixed by the conditions $\phi(0) = 2\phi_s(X) = \psi$. The successive $\phi(x)$ profiles versus θ is shown in Figure 20a.

The energy can be written as: $W(\psi) = E_s(1 - \cos(\psi/2))$. Over one period, $W(\psi)$ changes monotonously within $0 = W(0) \leq W(\psi) \leq W(2\pi) = 2E_s$. The remnant variational energy contains the pinning potential $V(\psi - \theta)$, which we take as $V(\phi) = V(1 - \cos\phi)$, and the energy of deformations $W(\psi)$ with $W(0) = 0$, $W(2\pi) = 2E_s$:

$$H(\psi, \theta) = V(\psi - \theta) + W(\psi) \quad (21)$$

The study of the extrema of this energy yields one or three solutions $\psi_i(\theta)$, $i = 1, 2, 3$ whose energies $E_i(\theta)$ are illustrated in Figure 20b. As an example, the profile $i = 3$

corresponds to a dilatation of the CDW wavelength at one side and a compression on the other side, in agreement with the observations [4,37]. The whole interval of θ, or some parts of it, can be either *mono-stable* or *bi-stable*. The last case corresponds to the coexistence of two locally stable branches: the *absolutely stable* one with the lower energy E_1 and the *metastable* one with a higher energy E_2. The same pair of branches can be regrouped also as the *ascending* branch E_+ for which $F_+(\theta) > 0$ and the *descending* one E_- with $F_-(\theta) < 0$, where E_a (with $a = \pm$) are the partial forces generated by the pinned state a. They correspond to the *retarded* and the *advanced* states at the impurity, respectively, and the two branches cross each other at $\theta = \pi$, with $\psi_\pm(\pi) < \pi$ (see Figure 20b). The barrier height, with respect to the metastable branch E_2, gives the activation energy for its decay: $U_b(\theta) = E_3(\theta) - E_2(\theta)$.

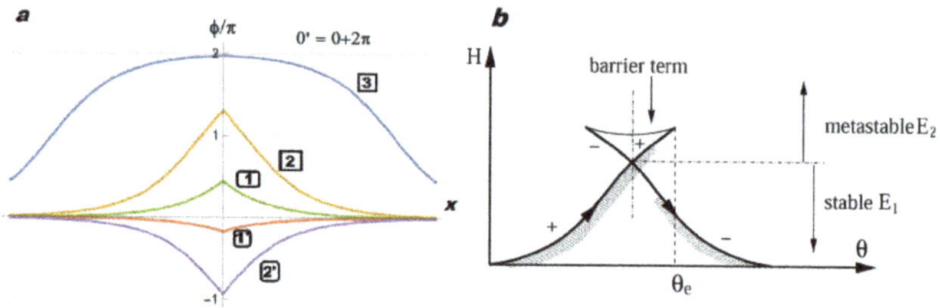

Figure 20. (**a**) The ϕ profile starts from the equilibrium position 0 with $\phi(x) \equiv 0$, and evolves through the shapes $1, 2, 3, 0'$. These configurations correspond to the retarded branch E_+ which becomes metastable after $\phi(0)$ crosses π. The phase ϕ will then follow the advanced profiles $1', 2', 3', 0$, corresponding to the branch E_-. If the retarded branch E_+ is less deformed, costing a smaller energy W, the relaxation $E_+ \to E_-$ is avoided, and the new development starts with the profile $0' = 0 + 2\pi$ corresponding to the infinitely divergent pair of solitons. (**b**) Energy branches for a bistable impurity. The upper line shows the barrier branch E_3. Solid lines show the locally stable branches $E_\pm$, also classified as $E_2 > E_1$. The difference $\Delta E = E_2 - E_1$ gives the dissipated energy. The difference $U = E_3 - E_2$ gives the activation energy for a decay of the metastable state E_2.

Let us consider the stationary process when the CDW moves with a constant phase velocity $\omega = -\dot{\phi} = \dot{\theta} = const$. The pinning force can be written as a weighted distribution of instantaneous forces:

$$F_{pin} = n_i \int_{\pi}^{\theta_{max}} d\theta F(\theta) \exp\left(- \int_{\pi}^{\theta} \frac{d\theta_1}{\omega\tau(\theta_1)} \right), \; F = \frac{d}{d\theta}\frac{\Delta E}{2} = \frac{F_+ - F_-}{2} \tag{22}$$

where the expression in the exponent generalizes, for a variable relaxation time $\tau(\theta)$, the natural guess for the decay probability as $\exp(-T_{slide}/\tau)$ where $T_{slide} = 2\pi/\omega$ is the period of the CDW sliding over the impurity site.

We shall limit the discussion to small velocities $\omega \ll \tau_\pi^{-1}$ where $\tau_\pi = \tau(\pi) \sim \exp(U_\pi/T)$, $U_\pi = U(\pi)$, is the maximal relaxation time in the region of the branch crossing the point $\theta = \pi$. The main contribution comes from the close vicinity of π: $\theta \approx \pi + \delta\theta$ where $\delta\theta \sim \omega\tau_\pi$. We can distinguish between two sub-regimes.

1. *Very small velocities*: with $\omega \ll \omega_\pi = T/(\tau_\pi F_\pi) \ll \tau_\pi^{-1}$ and $F_\pi = F(\pi)$. The decay happens as soon as the branch becomes metastable in a vicinity of π, even before the θ dependence is seen. The life time interval is $\delta\theta \sim v\tau_\pi$, hence Equation (22) yields the expression

$$F_{pin} = n_i \omega F_\pi \tau_\pi \tag{23}$$

which gives the phenomenological viscosity in the regime of the *linear collective conductivity*. It shows an activated behavior via τ_π^{-1} which can emulate the normal conductivity via thermally activated quasi-particles.

2. *Moderately small velocities*: with $\omega_\pi \ll \omega \ll \tau_\pi^{-1}$ and

$$F_{pin} \sim n_i T \ln[\omega\tau_\pi] \quad \text{i.e.,} \quad \omega \sim \tau_\pi^{-1} \exp(f/n_i T) \tag{24}$$

Convenient interpolation formulas for the two cases 1,2 can be obtained:

$$F_{pin} \approx T n_i \ln\left(1 + \omega \frac{\tau_\pi F_\pi}{2\pi T}\right); \quad \omega = \frac{2\pi T}{\tau_\pi F_\pi}\left(\exp \frac{F_{pin}}{T n_i} - 1\right) \tag{25}$$

The physics of $F_{pin} \sim \omega$ regime is given by the high probability to stay with the metastable branch in the course of small displacements $\delta\theta \sim \tau_\pi v$. The $F_{pin} \sim \ln \omega$ regime appears because that at higher ω a wider region of $\delta\theta$ is explored and the metastable branch starts to feel the decrease of the barrier (long in advance, there is either the termination point θ_e or the minimal barrier point θ_m), even if still unreachable at these moderate ω. The complete range of velocities was described in [69] and compared with experiments in [71].

5. Conclusions

This article summarizes different studies that have made it possible to observe defects in electronic crystals, such as charge and spin density waves, thanks to cutting-edge X-ray diffraction techniques and more particularly coherent X-ray diffraction. New information about the CDW static and dynamical regimes could be obtained thanks to the sensitivity of coherent X-rays to phase variations, especially in the sliding CDW state. Two main features emerge from this series of experiments: the CDW deformation, mainly due to pinning by sample surfaces and the propagation of a periodic lattice of phase shifts on top of the CDW. These two characteristics of sliding CDW are linked. The collective transport of charges in CDW systems is based on the dynamics of phase shifts induced by CDW deformations. After a threshold strain, the CDW elastic energy is released by creating a 2π soliton. Once created, the soliton propagates freely across the sample. Since the applied current is continuous, the creation of soliton is periodic, generating a soliton lattice in motion.

There is, however, still a lot to understand. The main difficulty lies in the fact that each system, despite identical aspects, has different properties. The diversity of measured responses probably shows that the type of transport is not identical depending on the system. Ultrashort and coherent X-ray pulses as the ones delivered by X-ray Free Electron Lasers could open new perspectives thanks to unprecedented spatial and temporal resolution to get a better understanding of this collective motion of charges.

Author Contributions: XRD data acquisition, participation in experiments and data analysis, E.B., D.L.B. and V.L.R.J.; theoretical work, N.K.; writing—original draft preparation, V.L.R.J., D.L.B. and N.K.; writing—review and editing, D.L.B. and N.K. All authors have read and agreed to the published version of the manuscript.

Funding: This research received no external funding.

Institutional Review Board Statement: Not applicable.

Informed Consent Statement: Not applicable.

Data Availability Statement: Data are available upon request to the corresponding author.

Acknowledgments: The work summarized in this review spans over 20 years and involves a large number of researchers. Without being able to name them all, we would first like to thank the PhD students who played an essential role in these works, E. Pinsolle and A. Rojo-Bravo. We would also like to give special thanks to A. Sinchenko whose mastery of resistivity and sample preparation techniques has always been the basis of all these experiments. We also thank all the collaborators who accompanied us (S. Ravy, C. Laulhé), and more particularly P. Monceau for his precious advice and for providing us with samples and, from a theoretical point of view, S. Brazovskii, especially for the last section of this review.

Conflicts of Interest: The authors declare no conflict of interest.

References

1. Lee, P.; Rice, T.; Anderson, P. Conductivity from charge or spin density waves. *Solid State Commun.* **1974**, *14*, 703–709. [CrossRef]
2. Fukuyama, H.; Lee, P.A. Dynamics of the charge-density wave. I. Impurity pinning in a single chain. *Phys. Rev. B* **1978**, *17*, 535–541. [CrossRef]
3. Ong, N.P.; Maki, K. Generation of charge-density-wave conduction noise by moving phase vortices. *Phys. Rev. B* **1985**, *32*, 6582–6590. [CrossRef] [PubMed]
4. DiCarlo, D.; Sweetland, E.; Sutton, M.; Brock, J.D.; Thorne, R.E. Field-induced charge-density-wave deformations and phase slip in $NbSe_3$. *Phys. Rev. Lett.* **1993**, *70*, 845–848. [CrossRef] [PubMed]
5. Rideau, D.; Monceau, P.; Currat, R.; Requardt, H.; Nad, F.; Lorenzo, J.E.; Brazovskii, S.; Detlefs, C.; Grübel, G. X-ray scattering evidence for macroscopic strong pinning centers in the sliding CDW state of NbSe 3. *Europhys. Lett.* **2002**, *57*, 471. [CrossRef]
6. Feinberg, D.; Friedel, J. Elastic and plastic deformations of charge density waves. *J. Phys. France* **1988**, *49*, 485–496. [CrossRef]
7. Sutton, M.; Mochrie, S.G.J.; Greytak, T.; Nagler, S.E.; Berman, L.E.; Held, G.A.; Stephenson, G.B. Observation of speckle by diffraction with coherent X-rays. *Nature* **1991**, *352*, 608–610. [CrossRef]
8. Le Bolloc'h, D.; Livet, F.; Bley, F.; Schulli, T.; Veron, M.; Metzger, T.H. X-ray diffraction from rectangular slits. *J. Synchrotron Radiat.* **2002**, *9*, 258–265. [CrossRef]
9. Jacques, V.L.R.; Le Bolloc'h, D.; Pinsolle, E.; Picca, F.E.; Ravy, S. Estimation of coherence properties of an undulator-generated X-ray beam from near-field and far-field slit diffraction visibilities. *Phys. Rev. B* **2012**, *86*, 144117. [CrossRef]
10. Le Bolloc'h, D.; Sadoc, J.F. X-ray coherent diffraction interpreted through the fractional Fourier transform. *Eur. Phys. J. B* **2011**, *81*, 481–487. [CrossRef]
11. Robinson, I.; Harder, R. Coherent X-ray diffraction imaging of strain at the nanoscale. *Nat. Mater.* **2009**, *8*, 291–298. [CrossRef]
12. Chapman, H.N.; Nugent, K.A. Coherent lensless X-ray imaging. *Nat. Photonics* **2010**, *4*, 833–839. [CrossRef]
13. Jacques, V.L.R.; Ravy, S.; Le Bolloc'h, D.; Pinsolle, E.; Sauvage-Simkin, M.; Livet, F. Bulk Dislocation Core Dissociation Probed by Coherent X Rays in Silicon. *Phys. Rev. Lett.* **2011**, *106*, 065502. [CrossRef]
14. Bolloc'h, D.L.; Ravy, S.; Dumas, J.; Marcus, J.; Livet, F.; Detlefs, C.; Yakhou, F.; Paolasini, L. Charge Density Wave Dislocation as Revealed by Coherent X-Ray Diffraction. *Phys. Rev. Lett.* **2005**, *95*, 116401. [CrossRef]
15. Jacques, V.L.; Bolloc'h, D.L.; Ravy, S.; Giles, C.; Livet, F.; Wilkins, S.B. Spin density wave dislocation in chromium probed by coherent X-ray diffraction. *Eur. Phys. J. B* **2009**, *70*, 317–325. [CrossRef]
16. Cowan, W. Strain and the spin-flip transition in chromium: Laudau theory. *J. Phys. F Met. Phys.* **1978**, *8*, 423. [CrossRef]
17. Young, C.; Sokoloff, J. The role of harmonics in the first order antiferromagnetic to paramagnetic transition in order. *J. Phys. F Met. Phys.* **1974**, *4*, 1304. [CrossRef]
18. Jacques, V.L.R.; Pinsolle, E.; Ravy, S.; Abramovici, G.; Le Bolloc'h, D. Charge- and spin-density waves observed through their spatial fluctuations by coherent and simultaneous X-ray diffraction. *Phys. Rev. B* **2014**, *89*, 245127. [CrossRef]
19. Gork'ov, L.; Grüner, G. *Charge Density Waves in Solids*; Elsevier Science: Amsterdam, The Netherlands, 1989.
20. Monceau, P. Electronic crystals: An experimental overview. *Adv. Phys.* **2012**, *61*, 325–581. [CrossRef]
21. Fleming, R.; Grimes, C. Sliding-Mode Conductivity in $NbSe_3$: Observation of a Threshold Electric Field and Conduction Noise. *Phys. Rev. Lett.* **1979**, *42*, 1423–1426. [CrossRef]
22. Thorne, R.E.; Lyons, W.G.; Lyding, J.W.; Tucker, J.R.; Bardeen, J. Charge-density-wave transport in quasi-one-dimensional conductors. I. Current oscillations. *Phys. Rev. B* **1987**, *35*, 6348–6359. [CrossRef] [PubMed]
23. Gruner, G. *Density Waves In Solids*; CRC Press: Boca Raton, FL, USA, 2018.
24. Frohlich, H. On the Theory of Superconductivity: The One-Dimensional Case. *Proc. R. Soc. A Math. Phys. Eng. Sci.* **1954**, *223*, 296–305. [CrossRef]
25. Aubry, S.; Quemerais, P. *Low-Dimensional Electronic Properties of Molybdenum Bronzes and Oxides*; Schlenker, C., Ed.; Springer: Dordrecht, The Netherlands, 1989; pp. 295–405.
26. Bardeen, J. Theory of Non-Ohmic Conduction from Charge-Density Waves in $NbSe_3$. *Phys. Rev. Lett.* **1979**, *42*, 1498–1500. [CrossRef]
27. Gor'kov, L.P. Boundary conditions and generation of periodic noise by a space-charge wave. *Sov. J. Exp. Theor. Phys. Lett.* **1983**, *38*, 87.
28. Gor'kov, L.P. Generation of oscillations by a running charge density wave. *Sov. J. Exp. Theor. Phys.* **1984**, *59*, 1818–1830.
29. Batistić, I.; Bjeliš, A.; Gor'kov, L.P. Generation of the coherent pulses by the CDW-motion. Solutions of the microscopic model equations. *J. Phys.* **1984**, *45*, 1049–1059. [CrossRef]
30. Maher, M.P.; Adelman, T.L.; DiCarlo, D.A.; McCarten, J.P.; Thorne, R.E. Charge-density-wave phase slip and contact effects in $NbSe_3$. *Phys. Rev. B* **1995**, *52*, 13850–13864. [CrossRef]
31. Maki, K. Quantum phase slip in charge and spin density waves. *Phys. Lett. A* **1995**, *202*, 313–316. [CrossRef]
32. Le Bolloc'h, D.; Jacques, V.L.R.; Kirova, N.; Dumas, J.; Ravy, S.; Marcus, J.; Livet, F. Observation of Correlations Up To the Micrometer Scale in Sliding Charge-Density Waves. *Phys. Rev. Lett.* **2008**, *100*, 096403. [CrossRef]
33. Jacques, V.L.R.; Le Bolloc'h, D.; Ravy, S.; Dumas, J.; Colin, C.V.; Mazzoli, C. Evolution of a large-periodicity soliton lattice in a current-driven electronic crystal. *Phys. Rev. B* **2012**, *85*, 035113. [CrossRef]
34. Sinchenko, A.A.; Lejay, P.; Monceau, P. Sliding charge-density wave in two-dimensional rare-earth tellurides. *Phys. Rev. B* **2012**, *85*, 241104. [CrossRef]

35. Le Bolloc'h, D.; Sinchenko, A.A.; Jacques, V.L.R.; Ortega, L.; Lorenzo, J.E.; Chahine, G.A.; Lejay, P.; Monceau, P. Effect of dimensionality on sliding charge density waves: The quasi-two-dimensional $TbTe_3$ system probed by coherent X-ray diffraction. *Phys. Rev. B* **2016**, *93*, 165124. [CrossRef]
36. Pinsolle, E.; Kirova, N.; Jacques, V.L.R.; Sinchenko, A.A.; Le Bolloc'h, D. Creep, Flow, and Phase Slippage Regimes: An Extensive View of the Sliding Charge-Density Wave Revealed by Coherent X-ray Diffraction. *Phys. Rev. Lett.* **2012**, *109*, 256402. [CrossRef]
37. Requardt, H.; Nad, F.Y.; Monceau, P.; Currat, R.; Lorenzo, J.E.; Brazovskii, S.; Kirova, N.; Grübel, G.; Vettier, C. Direct Observation of Charge Density Wave Current Conversion by Spatially Resolved Synchrotron X-Ray Studies in $NbSe_3$. *Phys. Rev. Lett.* **1998**, *80*, 5631–5634. [CrossRef]
38. Lemay, S.G.; de Lind van Wijngaarden, M.C.; Adelman, T.L.; Thorne, R.E. Spatial distribution of charge-density-wave phase slip in $NbSe_3$. *Phys. Rev. B* **1998**, *57*, 12781–12791. [CrossRef]
39. Bellec, E.; Gonzalez-Vallejo, I.; Jacques, V.L.R.; Sinchenko, A.A.; Orlov, A.P.; Monceau, P.; Leake, S.J.; Bolloc'h, D.L. Evidence of Charge Density Wave transverse pinning by X-ray micro-diffraction. *Phys. Rev. B* **2020**, *101*, 125122. [CrossRef]
40. Prester, M. Size effect in $NbSe_3$: Length dependence of the threshold field. *Phys. Rev. B* **1985**, *32*, 2621–2624. [CrossRef]
41. Yetman, P.; Gill, J. Size-dependent threshold fields for Fröhlich conduction in niobium triselenide: Possible evidence for pinning by the crystal surface. *Solid State Commun.* **1987**, *62*, 201–206. [CrossRef]
42. Mihály, G.; Hutiray, G.; Mihály, L. Local distortion of pinned charge density waves in orthorombic TaS3. *Solid State Commun.* **1983**, *48*, 203–205. [CrossRef]
43. McCarten, J.; DiCarlo, D.A.; Maher, M.P.; Adelman, T.L.; Thorne, R.E. Charge-density-wave pinning and finite-size effects in $NbSe_3$. *Phys. Rev. B* **1992**, *46*, 4456–4482. [CrossRef]
44. Borodin, D.; Nad', F.; Savitskaya, Y.; Zaitsev-Zotov, S. Nonlinear effects in small o-TaS3 samples. *Phys. B+C* **1986**, *143*, 73–75. [CrossRef]
45. Brazovskii, S.; Kirova, N.; Requardt, H.; Nad, F.Y.; Monceau, P.; Currat, R.; Lorenzo, J.E.; Grübel, G.; Vettier, C. Plastic sliding of charge density waves: X-ray space resolved-studies versus theory of current conversion. *Phys. Rev. B* **2000**, *61*, 10640–10650. [CrossRef]
46. Duan, J.M. Homogeneous quantum phase slippage in bulk charge-density-wave systems. *Phys. Rev. B* **1993**, *48*, 4860–4863. [CrossRef] [PubMed]
47. Hayashi, M.; Yoshioka, H. On the Ginzburg-Landau Free Energy of Charge Density Waves with a Three-Dimensional Order. *arXiv* **2000**, arXiv:cond-mat/0010102. [CrossRef]
48. Bellec, E. Study of Charge Density Wave Materials under Current by X-ray Diffraction. Ph.D. Thesis, Université Paris-Saclay, Paris, France, 2019.
49. Bellec, E.; Jacques, V.L.; Caillaux, J.; Le Bolloc'h, D. The essential role of surface pinning in the dynamics of charge density waves submitted to external dc fields. *Eur. Phys. J. B* **2020**, *93*, 165. [CrossRef]
50. Rojo-Bravo, A.; Jacques, V.L.R.; Le Bolloc'h, D.Collective transport of charges in charge density wave systems based on traveling soliton lattices. *Phys. Rev. B* **2016**, *94*, 201120. [CrossRef]
51. Fogel, M.B.; Trullinger, S.E.; Bishop, A.R.; Krumhansl, J.A. Dynamics of sine-Gordon solitons in the presence of perturbations. *Phys. Rev. B* **1977**, *15*, 1578–1592. [CrossRef]
52. Maher, M.P.; Ramakrishna, S.; Di Carlo, D.A.; Adelman, T.L.; Ambegaokar, V.; Brock, J.D.; Thorne, R.E. Charge-density-wave phase slip in NbSe3. *J. Phys. IV France* **1993**, *3*, C2-171–C2-174. [CrossRef]
53. Gill, J.C. Charge-density wave phase-slip in niobium triselenide : Dislocations and the growth of an, electronic crystal. *J. Phys. IV France* **1993**, *3*, C2-165–C2-170. [CrossRef]
54. Adelman, T.L.; de Lind van Wijngaarden, M.C.; Zaitsev-Zotov, S.V.; DiCarlo, D.; Thorne, R.E. Phase slip and the spatiotemporal response of charge-density waves in $NbSe_3$. *Phys. Rev. B* **1995**, *52*, R5483–R5486. [CrossRef]
55. Itkis, M.E.; Emerling, B.M.; Brill, J.W. Electrooptical imaging of charge-density wave phase gradients: Polarity and temperature dependence of phase slip. *Synth. Met.* **1997**, *86*, 1959–1960. [CrossRef]
56. Brun, C.; Wang, Z.Z.; Monceau, P.; Brazovskii, S. Surface Charge Density Wave Phase Transition in NbSe 3. *Phys. Rev. Lett.* **2010**, *104*, 256403. [CrossRef]
57. Isakovic, A.F.; Evans, P.G.; Kmetko, J.; Cicak, K.; Cai, Z.; Lai, B.; Thorne, R.E. Shear Modulus and Plasticity of a Driven Charge Density Wave. *Phys. Rev. Lett.* **2006**, *96*, 046401. [CrossRef]
58. Latyshev, Y.I.; Monceau, P.; Brazovskii, S.; Orlov, A.P.; Fournier, T. Subgap Collective Tunneling and Its Staircase Structure in Charge Density Waves. *Phys. Rev. Lett.* **2006**, *96*, 116402. [CrossRef]
59. Brazovskii, S.; Brun, C.; Wang, Z.Z.; Monceau, P. Scanning-Tunneling Microscope Imaging of Single-Electron Solitons in a Material with Incommensurate Charge-Density Waves. *Phys. Rev. Lett.* **2012**, *108*, 096801. [CrossRef]
60. Kim, T.H.; Yeom, H.W. Topological Solitons versus Nonsolitonic Phase Defects in a Quasi-One-Dimensional Charge-Density Wave. *Phys. Rev. Lett.* **2012**, *109*, 246802. [CrossRef]
61. Karpov, P.; Brazovskii, S. Pattern Formation and Aggregation in Ensembles of Solitons in Quasi One-Dimensional Electronic Systems. *Symmetry* **2022**, *14*, 972. [CrossRef]
62. Ravy, S.; Rouzière, S.; Pouget, J.P.; Brazovskii, S.; Marcus, J.; Bérar, J.F.; Elkaim, E. Disorder effects on the charge-density waves structure in V- and W-doped blue bronzes: Friedel oscillations and charge-density wave pinning. *Phys. Rev. B* **2006**, *74*, 174102. [CrossRef]

63. Noguera, C.; Pouget, J.P. Temperature dependence of the Peierls wavevector in quasi one dimensional conductors. *J. Phys. I* **1991**, *1*, 1035. [CrossRef]
64. Artemenko, S.N.; Ya, V.; Pokrovskii, S.Z.Z. Electron-hole balance and semiconductor properties of quasi-one-dimensional charge density-wave conductors. *Sov. Phys. JETP* **1996**, *110*.
65. Pouget, J.P.; Hennion, B.; Escribe-Filippini, C.; Sato, M. Neutron-scattering investigations of the Kohn anomaly and of the phase and amplitude charge-density-wave excitations of the blue bronze $K_{0.3}MoO_3$. *Phys. Rev. B* **1991**, *43*, 8421–8430. [CrossRef] [PubMed]
66. Hennion, B.; Pouget, J.P.; Sato, M. Charge-density-wave phase elasticity of the blue bronze. *Phys. Rev. Lett.* **1992**, *68*, 2374–2377. [CrossRef] [PubMed]
67. Ravy, S.; Requardt, H.; Le Bolloc'h, D.; Foury-Leylekian, P.; Pouget, J.P.; Currat, R.; Monceau, P.; Krisch, M. Inelastic X-ray scattering study of charge-density-wave dynamics in the $Rb_{0.3}MoO_3$ blue bronze. *Phys. Rev. B* **2004**, *69*, 115113. [CrossRef]
68. Brazovskii, S.; Kirova, N. From chiral anomaly to two-fluid hydrodynamics for electronic vortices. *Ann. Phys.* **2019**, *403*, 184–197. [CrossRef]
69. Brazovskii, S.; Nattermann, T. Pinning and sliding of driven elastic systems: From domain walls to charge density waves. *Adv. Phys.* **2004**, *53*, 177–252. [CrossRef]
70. Zybtsev, S.G.; Pokrovskii, V.Y. Quantization of states and strain-induced transformation of charge-density waves in the quasi-one-dimensional conductor TaS_3. *Phys. Rev. B* **2016**, *94*, 115140. [CrossRef]
71. Ogawa, N.; Miyano, K.; Brazovski, S. Optical excitation in the creep phase of plastic charge-density waves. *Phys. Rev. B* **2005**, *71*, 075118. [CrossRef]

symmetry

MDPI

Article

Pattern Formation and Aggregation in Ensembles of Solitons in Quasi One-Dimensional Electronic Systems

Petr Karpov [1,2,*] and Serguei Brazovskii [3]

1 Max Planck Institute of Physics of Complex Systems, D-01187 Dresden, Germany
2 Arnold Sommerfeld Center for Theoretical Physics, Ludwig-Maximilians University, D-80539 Munchen, Germany
3 LPTMS, CNRS, University Paris-Saclay, 91405 Orsay, France; serguei.brazovski@universite-paris-saclay.fr
* Correspondence: karpov@pks.mpg.de

Abstract: Broken symmetries of quasi one-dimensional electronic systems give rise to microscopic solitons taking roles of carriers of the charge or spin. The double degeneracy gives rise to solitons as kinks of the scalar order parameter A; the continuous degeneracy for the complex order parameter $A\exp(i\theta)$ gives rise to phase vortices, amplitudes solitons, and their combinations. These degrees of freedom can be controlled or accessed independently via either the spin polarization or the charge doping. The long-range ordering in dimensions above one imposes super-long-range confinement forces upon the solitons, leading to a sequence of phase transitions in their ensembles. The higher-temperature T transition enforces the confinement of solitons into topologically bound complexes: pairs of kinks or the amplitude solitons dressed by exotic half-integer vortices. At a second lower T transition, the solitons aggregate into rods of bi-kinks or into walls of amplitude solitons terminated by rings of half-integer vortices. With lowering T, the walls multiply, passing sequentially across the sample. Here, we summarize results of a numerical modeling for different symmetries, for charged and neutral soliton, in two and three dimensions. The efficient Monte Carlo algorithm, preserving the number of solitons, was employed which substantially facilitates the calculations, allowing to extend them to the three-dimensional case and to include the long-range Coulomb interactions.

Keywords: topological defect; soliton; stripe; confinement; aggregation

1. Introduction: Confinement and Phase Transitions in Ensembles of Solitons

Strongly correlated electronic systems are prone to spontaneous symmetry breaking which brings about a manifold of degenerate ground states. Such degeneracy makes possible the existence of topologically non-trivial configurations connecting different states (see [1] for a latest review). These include, for example, domain walls and stripes, dislocations and vortex lines, as well as their complex aggregates. In two- and three-dimensional systems such objects are still macroscopic. However, in one-dimensional (1D) systems they become truly microscopic objects carrying energy, charge, and other quantum numbers of a single-electron scale. It appears, frequently, that such microscopic quasi-particles, called solitons, become the lowest energy excitations and take over the role of conventional electrons, e.g., in transport or optical properties (see [2,3] for short reviews). Since the solitons possess such non-zero quantum numbers as charge or spin, their stable ensembles can be controllably created and/or observed by the usual means, such as scanning tunneling microscopy. Such collective systems typically experience a sequence of phase transitions, accompanied by the formation of structures at an increasing scale: from individual solitons, via their microscopic complexes and growing aggregates to macroscopic domain walls and stripes.

The popular notion of "topological solitons" is a shortening for "topologically nontrivial solitons" which still needs precision such as "topologically stable ..." or "topologically protected ...". The "topological stability" is based on conservation of "topological charges" and the related "irreducibility of trajectories" connecting different states of a systems, see reviews [4,5]. Following only the minimal feature of a spanning trajectory, with no respect to its reducibility, we shall imply a broader while less precise definition of a "topologically nontrivial soliton" as a local configuration exploring a manifold of degenerate ground state connecting different equivalent ones. The configurations may not be topologically stable, thus allowing for their transmutations with trivial electronic excitations. Such solitons appear or are preserved because they are energetically preferable and/or because the total electric charge or spin, which these solitons carry on, are monitored externally.

A major puzzle, as well as the inspiration, is that experimentally the solitons were observed or looked for within the low-temperature phases with a long-range order. The commonly unattended obstacle is the effect of the confinement of solitons (see [2] and refs. therein). Commuting between degenerate minima at only one chain would lead to a loss of the interchain ordering energy proportional to $|x|$—the distance along the chain till the next soliton or a boundary. This energy dominates at long distances even if it can be unimportant locally for a weak interchain interaction.

These interactions can appear already in some specific $1D$ systems where the ground state degeneracy is not exact, so the soliton connects the true and the false vacuums losing the confinement energy. The effect of confinement is omnipresent at higher dimensions where the interchain interactions, which are responsible for establishing the long-range $2D$ or $3D$ ordering, lift the degeneracy locally. In cases of discrete symmetries (when solitons are the amplitude kinks as in the Z_2 case of the dimerization) the solitons are bound in topologically trivial pairs with an option for a subsequent phase transition to form cross-sample domain walls [6–8]. Much richer are the cases of continuous symmetries: the phase degeneracy in superconductors and incommensurate charge density waves (ICDW), the directional degeneracy in spin-isotropic antiferromagnets, and both phase and spin degeneracy in spin density waves (SDW). Here, the gapless mode can cure the interruption from the amplitude kink which allows for individual solitons to enter the low-temperature phases with long-range ordered states. The solitons adapt by forming topologically bound combined complexes with half-integer vortices of gapless modes: π-rotons [2,3]. For cases of repulsing and attracting electronic interactions correspondingly, that results in spin- or charge-roton configurations with charge- or spin-amplitude solitons localized in the core. Beyond the quasi-1D electronic systems, the interference of different types of topological defects has been noticed already in the theory of superfluid phases in surface layers of He_3 [9,10] and recently in doped antiferromagnets [11].

In this short review, we shall summarize and update our recent, largely numerical, studies on statistical ensembles of solitons [12,13], see also [14]. We refer to these publications for a broad review of experimental motivation and for a wider citations of the theoretical literature, while here we shall be concerned with only major theoretical aspects. Section 2 will be devoted to solitons in systems with a discrete double degeneracy. Separate sections will describe cases of neutral and charged solitons, modeling in $D = 2$ and $D = 3$ dimensions, slow a cooling and a quench. Section 3 will be devoted to a richer system with a combined amplitude-phase degeneracy, with numerical results shown for $D = 2$, prone to combined topological defects with half-integer vorticity. Section 4 will be devoted to discussions and conclusions.

2. Discrete Symmetry: Solitons in Doubly-Degenerate Systems

In this chapter we shall address the case of a discrete symmetry breaking, particularly the double degeneracy, which is a very common phenomenon of a dimerization of bonds (the family of Peierls-like transitions) and/or of sites (the family of transitions with charge ordering or disproportionation). The alternation of spontaneously different sites or bonds allows for disruptive defects which passing changes the sign of the order parameter $A \to -A$, hence a soliton with a nickname the "kink". In relevant electronic systems the soliton forms a potential wall possessing the bound "mid-gap" state which is allowed to capture one or two electrons, giving rise to oppositely charged spinless and/or neutral (or sometimes also charged) spin-carrying solitons. In a 1D-system or at high-temperature (T) disordered phase at a higher dimension (D), the randomly distributed solitons break the long range correlation at a length of the order of a mean distance among the kinks. Commonly, the on-chain interactions among solitons are short range (falling exponentially) repulsive. There are special cases of a highly degenerate energy landscape with a coalescence of different ground states, when the interaction among solitons becomes long ranged, decaying only as power laws (see [15] and refs. therein). Still, this is a weak effect in comparison with non-decaying confinement forces produced by interchain interactions which we shall study below.

At $D > 1$ with lowering T, the interchain correlation of kinks enters the game drastically. Figure 1a,b shows that the energy of interchain interactions increases with growing separation between two solitons both on the same chain or on neighboring chains. Hence, there are two tendencies: either towards the formation of intra-chain bound pairs of solitons (which we shall call bi-solitons or bi-kinks), or towards inter-chain aggregation of solitons into the kink walls. As a result of these competing trends, the system goes through two phase transitions: the intra-chain pairing of solitons at a higher $T = T_1$, see Figure 1c,d, and the inter-chain aggregation of pairs at a lower $T = T_2$, see Figure 1e,f. The upper T_1 is the transition temperature where the long range is established for the order parameter: $\langle A \rangle \neq 0$ at $T < T_1$, while for solitons it is the onset of their pairwise confinement. For a three-dimensional system, T_2 is a specific phase transition point; below this temperature there appear domain wall planes, which pass through the entire cross section of the sample. For a two-dimensional system, this is not a phase transition, but rather a crossover: with lowering the temperature below T_2, finite bi-solitonic rods (pieces of paired domain walls) gradually grow in the transverse direction. For finite systems, as in our numerical simulations, there exists a sample-dependent temperature T_F, at which the first wall crosses the entire sample even in a two-dimensional system.

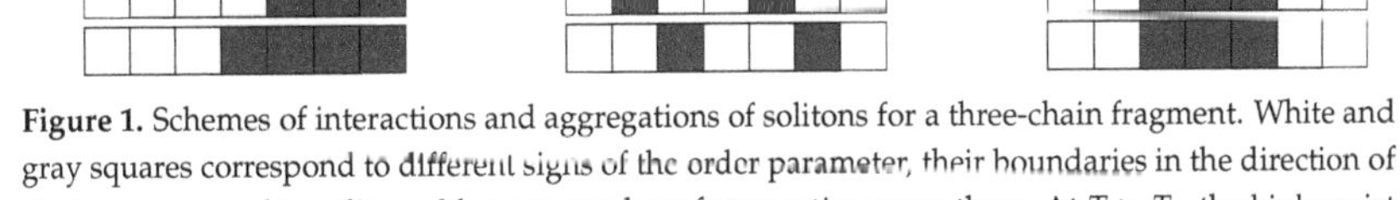

Figure 1. Schemes of interactions and aggregations of solitons for a three-chain fragment. White and gray squares correspond to different signs of the order parameter, their boundaries in the direction of chains correspond to solitons; blue arrows show forces acting upon them. At $T > T_1$, the kinks exist as separate objects, but they already experience a long-range attraction that links them into pairs on the same chain (**a**) or into walls across the chains (**b**). At $T_1 > T > T_2$, the solitons are bound forming an ensemble of bisolitons (**c**). The lengths of bisolitons are large and fluctuate at $T_1 > T > J_\perp$ (**c**). The bisolitonic pairs are tight at $J_\perp > T > T_2$ (**d**). At $T < T_2$, the bisolitons aggregate into growing bisolitonic walls (**e**) and then, reaching the sample boundaries, they split into isolated walls of kinks (**f**).

This qualitative picture is supported by an exact solution available for a $2D$ system of neutral solitons with some qualitative extensions to the $3D$ case [6]. The case of electrically charged solitons has been addressed in [7,8], but the numerical studies [7] suffered with a restrictive constraint: the bi-kink pairs were not allowed to jump between the chains. Following the approach of [12], here we shall present the results of unrestricted numerical modeling performed for the challenging case of a $3D$ system, both for neutral and charged ensembles of solitons.

2.1. Mapping to the Constrained Ising Model

The doubly-degenerate system can be mapped [6] upon the Ising model for spin variables $S_{n,\alpha} = \pm 1$ defined upon the discretized (with a distance $a_{||}$) sites n of chains α. The local density of kinks $\rho_{n,\alpha}$ can be expressed as

$$\rho_{n,\alpha} = (1 - S_{n,\alpha} S_{n+1,\alpha})/2\,,\ \langle \rho_{n,\alpha} \rangle = \nu = N/V,$$

where ν is the mean concentration of solitons per site, N is their total number, $V = L \times H^{D-1}$ is the system volume with L, H being the sizes along and transverse the chains. The mean density ν of solitons can be controlled by their chemical potential μ, then the Gibbs energy $\widetilde{H}_0$ for the grand canonical ensemble of solitons becomes

$$\widetilde{H}_0 = H_0 + (E_s - \mu)N = -J_{\perp} \sum_{\langle \alpha,\beta \rangle n} S_{n,\alpha} S_{n,\beta} - J_{||} \sum_{\alpha,n} (S_{n,\alpha} S_{n+1,\alpha} - 1), \tag{1}$$

where the transverse spin-coupling constant $J_{\perp}$ comes from the given energy of interchain interactions while the longitudinal one $J_{||} = (\mu - E_s)/2$, where E_s is the energy paid to create the defect, is variable together with μ.

For electrically charged solitons we must add also [7,8] the Coulomb energy H_C:

$$\widetilde{H} = \widetilde{H}_0 + H_C\,,\ H_C = \frac{e^2}{2\epsilon} \sum_{n,m;\alpha,\beta} \frac{(\rho_{n,\alpha} - \nu)(\rho_{m,\beta} - \nu)}{|\mathbf{r}_{n,\alpha} - \mathbf{r}_{m,\beta}|}, \tag{2}$$

where e is the electron charge and ϵ is the dielectric constant.

In the resulting anisotropic Ising model (1), only the interchain coupling constant $J_{\perp}$ has a physical nature and remains fixed, while the intrachain coupling constant $J_{||}$ is determined by the chemical potential of solitons. Controlling μ, a grand canonical ensemble of solitons in D dimensions can be mapped to the D-dimensional Ising model, whereas the canonical ensemble can be mapped to a stack of non-interacting $(D-1)$-dimensional Ising models with a global constraint $\langle \rho_{n,\alpha} \rangle = \nu$. The latter case of controlling the concentration ν corresponds to our physical situation of interest, where the source of the most interesting behavior comes from inverting the self-consistency condition $\nu(\mu, T)$ to $\mu(\nu, T)$, which means moving along a special line in the parameter space of the Ising model $\{J_{||}, J_{\perp}, T\}$. Alternatively, it is possible to construct a numerical procedure [12], which allows us to implement the constraint explicitly by preserving the total number of solitons. In this case, we work directly with the canonical ensemble and taking advantage of working with $(D-1)$-dimensional systems in which only physical interchain interaction is present.

2.2. Summary for the Effective Ising Model for a Neutral System

Recall the picture of equilibrium states for a neutral system (Figure 1). As the temperature decreases at a given concentration ν of solitons, the system goes through a sequence of two phase transitions (at $T = T_1, T_2$) and a crossover between them (at $T \approx J_{\perp}$). These temperatures are governed by three energy scales: $J_{\perp}$ as the local energy of interchain ordering, larger $T_1 \propto J_{\perp}/\nu$ as the nonlocal energy of soliton confinement, and the smaller $T_2 \propto J_{\perp}/\ln(1/\nu)$ as characteristic energy of the transverse aggregation of solitons.

I. $T_1 \propto J_{\perp}/\nu$. For the Ising order parameter $\langle S \rangle$, this is simply the temperature $T_1 = T_c$ of the second-order phase transition to a state with its non-zero mean at $T < T_c$. However, for an ensemble of solitons, this is a confinement–deconfinement phase transition. It occurs when the temperature drops below the average interchain interaction Fl, where $l = a_{||}/\nu$ is the mean distance between solitons. At $T < T_1$, individual solitons become confined into loose intra-chain pairs; at the lower $T \sim J_{\perp}$, the pairs evolve into tightly bound "bisolitons".

II. $T_2 \propto J_{\perp}/\ln(1/\nu) < J_{\perp}$. This is the temperature of solitons' transverse aggregation, when the first domain walls crossing the entire sample emerge. Since, at $T = T_2$, we have $J_{||}(T_2) = 0$, the effective dimensionality of the system D is reduced by 1. With aggregation of bisolitons into the domain walls, the confinement energy is gained due to the proper ordering between neighboring chains. With that, the entropy is lost, and this balance sets the temperature T_2.

The bisolitonic pairs still coexist with walls below T_2, but as the temperature decreases further, the pairs disappear, giving material for building new domain walls. The basic intuition for ordered systems (e.g., Ising model) tells us that at low T the system remains in a simple form of a single ordered domain, diluted by rare spin-flips—tightly-bound bisolitons with an energy of $2NJ_\perp$. However, with our constraints, the picture is different. The self-consistent concentration of bi-kinks is

$$\nu_{bk} \approx \exp\left(-\frac{4J_{||} + 2NJ_\perp}{T}\right), \tag{3}$$

where N is the number of nearest chains. Since the concentration is fixed at $\nu_{bk} = \nu/2$, then the decrease in temperature must be compensated by the decrease in $J_{||}$, which nevertheless must remain non-negative $J_{||} \geq 0$. Hence, a new soliton storage reservoir must be opened, with $J_{||}$ remaining equal to zero when T decreases below $T_2(\nu)$ such that $\nu = 2\exp(-2NJ_\perp/T_2)$, i.e.,

$$T_2 \approx \frac{2NJ_\perp}{\ln(2/\nu)}. \tag{4}$$

This reservoir is a system of stripes (lines in $2D$ or planes in $3D$) that cross the entire sample, dividing it into domains of alternating magnetization. In three-dimensional systems, below T_2 the transverse layers do not interact and $J_{||}$ stays at zero value, hence T_2 must be indeed a sharp phase transition. In two-dimensional systems, T_2 is only a crossover for nucleation of growing rods of finite length. In $2D$, only for a finite sample of a width H, the rods grow through its entire cross-section at a temperature $T_F \approx 4J_\perp / \ln(H/2\nu)$.

2.3. *Numerical Approach for Ensembles of Amplitude Solitons*

In this section, we review our numerical Monte Carlo (MC) results for $2D$ and $3D$ cases of neutral and charged solitons [12]. As we pointed out above, in the numerical simulations it is more efficient to preserve a fixed number of solitons, while for analytical calculations it is more favorable to use self-consistent $J_{||}(T, \nu)$ approach, which was more convenient for the qualitative analysis and the access to exact solutions in $D = 2$ [6]. We will work in the canonical ensemble representation, keeping fixed only the total number of solitons, while allowing to exchange them between the chains (in contrast to the more restrictive condition of fixing their number at the individual chains used in Ref. [7]). For charged solitons, it is also necessary to consider the long-range Coulomb interactions, which is always a difficult task, especially in our case of pattern-prone system, where locally non-compensated charges appear at growing scales.

2.3.1. Neutral Solitons

We start with the basic case of the ensemble of neutral solitons. Figure 2 presents typical patterns of formation and reproduction of domain walls and discs (paired walls). At higher temperatures ($T = 2.1J_\perp > T_2$, Figure 2a) the system is in an ordered state diluted by a gas of bisolitons. With the formation of the first pair of domain walls (at $T = 2.0J_\perp \approx T_2$, Figure 2b), the concentration of non-condensed solitons drops, further decreasing gradually to cure the voids in already existing bisolitonic walls which then diverge and form two solitonic ones, giving rise to a domain of inverted spins between them. Then, the concentration of noncondensed solitons drops abruptly again with the formation of the another bisolitonic wall ($T = 1.6J_\perp < T_2$, Figure 2c).

Interestingly, the second bisolitonic wall emerges in the vicinity of the first soliton wall. Presumably, this is because the motion of bisolitons intending to build the second wall is forbidden in the direction of the existing wall, which halves the probability of their escape, and thus facilitates the aggregation of the incipient second pair of walls.

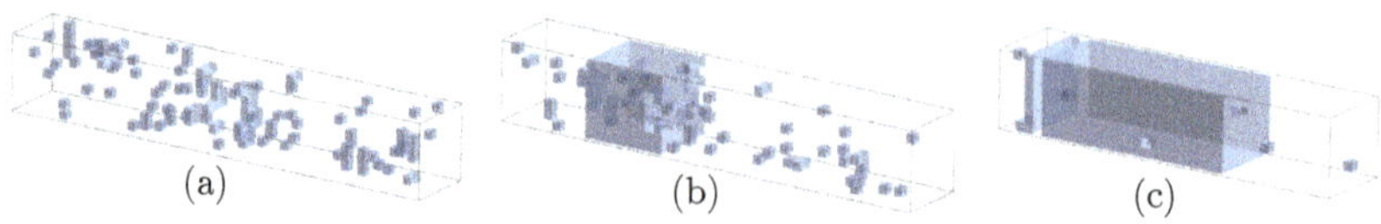

Figure 2. Formation of domain walls in a system of neutral solitons with the size $50 \times 8 \times 8$ and the solitons' concentration $\nu = 0.08$. (**a**) $T = 2.1J_{\perp}$—no walls; (**b**) $T = 2.0J_{\perp}$—two soliton walls; (**c**) $T = 1.6J_{\perp}$—four soliton walls.

2.3.2. Charged Solitons

Next, we address the case of electrically charged solions, focusing on the most interesting regime of intermediate values $J_{\perp}/H^2 \ll V_C \ll J_{\perp}$ of the Coulomb parameter $V_C = e^2/\epsilon a_{\perp}$ ($a_{\perp}$ is the interchain distance). Now, V_C is locally sufficiently weak, that it does not prevent the binding of kinks into bi-kinks; but even more than that—it does not prevent the initial aggregation of bi-kinks at neighboring chains into small disks. However, CI plays an important role for the large-scale structures, e.g., for large disks and domain walls, because for $V_C \gg J_{\perp}/H^2$ the formation of a wall planes is energetically unfavorable.

Thus, in contrast to the neutral solitons and weak CI cases, we do not observe a sharp transition of wall formation for intermediate values of the Coulomb parameter. We still observe finite bi-walls, possessing now the void defects. As the temperature approaches zero, these defects can merge, cutting the wall along one of the transverse directions (Figure 3a). At higher values of the Coulomb parameter, no plane walls are formed at all, but filamentous stripes emerge instead, which are infinite along one of the transverse directions and finite along the other one (Figure 3b).

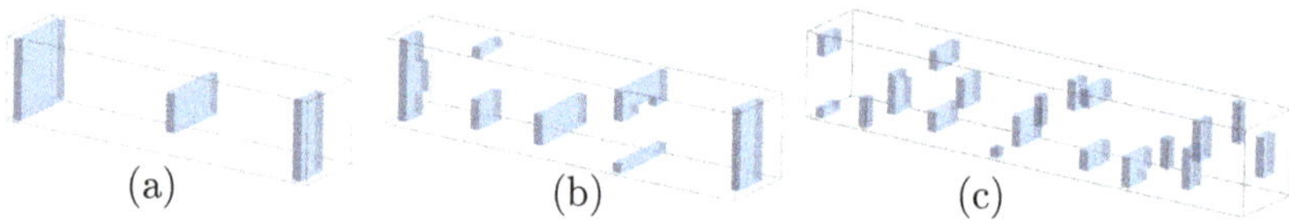

Figure 3. Disintegration of domain walls as the CI increases. Modeling for a system with a size $50 \times 8 \times 8$ at $\nu = 0.08$ and $T = 0.1J_{\perp}$ for different values of the Coulomb parameter: (**a**) $V_C = 0.02J_{\perp}$, (**b**) $V_C = 0.1J_{\perp}$, (**c**) $V_C = 0.3J_{\perp}$.

Formation of lines rather than planes demonstrates the extreme sensitivity of CIs to the pattern geometry. A charged plane wall would create a constant electric field $E = 4\pi e/(\epsilon a_{\perp}^2)$ in the chains' x direction, whose repulsive force eE would oppose directly the attractive confinement force $F = 2ZJ_{\perp}$ overpassing it at the intermediate CI, hence no stable planes could exist. However, forming the stripes (in one transverse y direction) at the expense of the part of the confinement energy (lost in the other transverse direction z), the system generates a decreasing electric field $E \propto 1/\sqrt{x^2+z^2}$ which falls below the confinement force at sufficiently large distances, hence preserving the partial aggregation. We expect that for an infinite system these structures will further disintegrate into finite pieces. For an even higher value of the Coulomb parameter, the formation of stripes becomes also energetically unfavorable even for a finite system, and we observe only disks of bi-kinks (Figure 3c). For an even stronger Coulomb interaction $V_C \gtrsim J_{\perp}$, the transverse disks shrink to the minimum size of one bi-kink which form a Wigner "liquid", then the crystal, and finally a crystal of solitons. The latter state is the universal limit of any system of dilute point charges.

3. Pattern Formation in Ensembles of Phase-Amplitude Solitons for a Complex Order Parameter

In this chapter, we construct and study the coarse-grained model of XY–Ising type, which describes the local and the long-range orderings under the created and then monitored mean concentration of solitons. We shall demonstrate an evolution of ensembles of solitons via a sequence of phase transitions or crossovers. At $T < T_1$ the interchain correlations settle in, and dilute amplitude solitons are dressed by half-integer vortex rings (vortex-antivortex (VA) pairs in $D = 2$). Upon further cooling towards a lower T_2, the solitons start to aggregate in the transverse direction forming domain walls terminated by the half-VA pairs. The terminating half-vortices at neighboring walls interact by pseudo-Coulomb forces. Attraction and subsequent annihilation of vortices of opposite signs belonging to neighboring walls favors the gluing of walls, either sequentially, elongating their lengths, or parallelly forming bisolitonic rods. Both scenarios will be visualized in our modeling. At $T < T_2$ in $D = 3$, or at a size-dependent temperature $T_F \sim T_2$ at $D = 2$, the solitonic walls grow across the whole sample and finally the regular "stripe phase" sets in.

3.1. Solitons in Quasi-1D Systems with Combined Amplitude and Phase Degeneracies

Most important cases of degenerate ground states in electronic systems (superconductivity, incommensurate CDWs) invoke a complex order parameter $\mathcal{O} = A\exp(i\theta)$. Almost commonly, the amplitude A is assumed to be positive $A > 0$ being associated with a "superfluid" or "collective" densities $A = \rho_s, \rho_c$ known also as the "phase rigidity". With respect to the flexible phase θ which variations give rise to observable collective currents, sound waves and vortices, the amplitude is considered to be only weakly perturbed, except for its enforced vanishing at cores of phase vortices. However, the microscopic insight at quasi-1D electronic systems (see [2,3]) shows that actually A is an independent local degree of freedom which should not be positively defined and, moreover, the alternations of its sign are related to monitored quantum numbers.

Recall briefly some most basic facts on solitons in spin-singlet ground states of electronic systems: ICDWs and quasi-1D superconductors (see refs. in [2,3]). The ICDW is a weak crystal $A\cos(2K_f x + \theta)$ of the electronic density and atomic displacements with the order parameter is $\mathcal{O}_{ICDW} = A\exp(i\theta)$. The singlet pair can be broken into spin 1/2 components, but instead of expectedly liberated electron–hole pair at the gap rims $\pm\Delta$, there will be spin-carrying "amplitude solitons" (AS)—nodes of the order parameter distributed over the length $\xi \sim 1/\Delta$. The total energy of the soliton is $E_s \approx 2\Delta/3 < \Delta$ making it energetically favorable with respect to the energy Δ of the undressed unpaired electron which now is trapped at the mid-gap state associated with the amplitude soliton. This electron brings the spin 1/2, but its charge is compensated to zero by the local dilatation in the occupied manifold of singlet vacuum states. That makes the AS a CDW realization of the "spinon", with a direct generalization to a superconductor [16] where the AS appears as the node line in the so called Fulde–Ferrel–Larkin–Ovchinnikov (FFLO) phase in spin-polarized superconductors (see [3] and refs. therein).

The amplitude soliton $A(x = -\infty) \Rightarrow -A(x = +\infty)$ performs the sign change of the order parameter $\mathcal{O}$ at an arbitrary phase $\theta = const$. As a non-trivial topological object (the order parameter $\mathcal{O}_{cdw}(x) = A(x)\exp(i\theta(x))$ does not map onto itself), the pure AS (with $\theta(x) = const$) is prohibited in $D > 1$ environment. At higher space (or space-time for instantons) dimensions D, the requirement always appears that the order parameter is mapped onto itself. Even having a lower energy in comparison with the band electron, the purely amplitude soliton cannot be created dynamically already in the $D = 1$ space dimension, and it is not allowed even stationary at $D > 1$. The conflict is resolved recalling the combined symmetry: the amplitude kink $A \Rightarrow -A$ accompanied by the half-integer $\theta \Rightarrow \theta \pm \pi$ vortex of the phase rotation, for which the factor $\exp(i\theta(x))$ compensates for the amplitude sign change. The resulting kink–roton complex allows for several interpretations in applications to ICDW or superconductors: $2D$ view is a pair of π-vortices sharing the common core which accommodates one unpaired spin which stabilizes the state; $3D$ view is a ring of a half-integer vortex ring which center confines the spin; at any $D > 1$ this is a nucleus of the melted FFLO phase in the spin-polarized superconductors or its CDW analogue, see [3]. In repulsive electronic systems, basically the SDW and the doped antiferromagnetic state, the situation is mirror symmetrical: the amplitude node carries the electronic charge $\pm e$ while the wings of half-integer vortices provide a distributed spin 1/2, see [2,3].

3.2. The Effective Model and the Qualitative Phase Diagram

A discrete description of a quasi-1D system with both the amplitude A and the phase θ degrees of freedom can be based upon the energy functional

$$H = -C\sum_r \cos(\theta_r - \theta_{r+\hat{x}}) - J_\perp \sum_r s_r s_{r+\hat{y}} \cos(\theta_r - \theta_{r+\hat{y}}) - J_{||} \sum_r s_r s_{r+\hat{x}}. \tag{5}$$

Here, the Ising variable $s_r = \pm 1$ corresponds to the normalized amplitude of the complex order parameter and the angle θ_r corresponds to its phase; r runs over sites of a lattice including the chains' direction $\hat{x}$ and the interchain vector $\hat{y}$. The last term with $J_{||} = (\mu - E_s)/2$ fixes the chemical potential μ of amplitude solitons, similarly to the Z_2 symmetry case of the previous chapter.

The Hamiltonian (5) describes two interfering subsystems: an Ising model which interchain coupling is modulated by phase distortions, and the XY model which interchain rigidity is interrupted by defects of the spin ordering. It is conceptually close to the Korshunov's 2θ model [9,10] and the coupled XY–Ising model [17,18]. The Hamiltonian (5) is invariant under a semi-local gauge transformation of simultaneous flipping of all Ising spins and all XY vectors within any given chain y independently: $s_r \to -s_r$, $\theta_r \to \theta_r + \pi$. That implies vanishing of macroscopical averages and the interchain correlations of Ising and XY fields separately: $\langle s_r \rangle = 0$, $\langle \exp(i\theta_r) \rangle = 0$. Still, a combined variable – the physical order parameter $\Delta_r = s_r e^{i\theta_r}$ – possesses the correlation functions and can acquire a nontrivial average.

Similarly to the case of the Z_2 symmetry, we expect two successive phase transitions at T_1 and T_2. The long range ordering is established at T_1, so that only at $T > T_1$ the amplitude solitons exist in their pure form at uncorrelated chains. For the $3D$ system, the upper transition temperature T_1 of the Z_2 case is generalized as a critical temperature T_c which character will change along the line $T_c(\nu)$, see Figure 4. The line can be determined within the mean-field approximation with the expansion in powers of the order parameter η:

$$\langle s_r \cos(\theta_r) \rangle = \eta \,, \text{ while } \langle s_r \rangle = 0 \,, \, \langle \cos(\theta_r) \rangle = 0.$$

At $T > T_1$, $\eta = 0$; the interchain interactions can be neglected so that only intra-chain correlations are present:

$$\langle s_{00} s_{0x} \cos(\theta_{00}) \cos(\theta_{0x}) \rangle \Rightarrow \langle s_{00} s_{0x} \rangle \langle \cos(\theta_{00}) \cos(\theta_{0x}) \rangle = \exp(-\nu|x| - |x|/l_T) \,, \, l_T \sim C/T,$$

where l_T is the phase correlation length in the 1D regime. The mean-field Hamiltonian

$$H = -C\sum_r \cos(\theta_r - \theta_{r+\hat{x}}) - J_\perp \eta \sum_r s_r \cos(\theta_r) - J_{||} \sum_r s_r s_{r+\hat{x}}$$

gives rise to the self-consistent equation which, in the first order in η, reads

$$\eta = \eta N J_\perp \sum_x \langle s_{00} s_{0x} \cos(\theta_{00}) \cos(\theta_{0x}) \rangle \,, \, \eta \neq 0 \text{ if :}$$

$$1 = N J_\perp / T_c \sum_x \exp(-\nu|x| - |x|/l_T) = \frac{2NJ_\perp/T_c}{\nu + T_c/C} \tag{6}$$

where N is the number of nearest interacting chains. Equation (6) gives the dependence $T_c(\nu)$ via the equation $2NJ_\perp/T_c - T_c/C = \nu$. There are two regimes:

$$T_c \approx NJ_\perp/\nu \text{ at } \nu > \nu^* = T_c/C \approx \sqrt{2NJ_\perp/C} \tag{7}$$
$$T_c \approx \sqrt{NJ_\perp C} \text{ at } \nu < \nu^*, \;\; \max(T_c) = T_{c0} = \sqrt{2NJ_\perp C} \text{ at } \nu = 0$$

with a crossover at $\nu = \nu^*$. The first line in Equation (7) implies a weak interchain coupling: $J_\perp \ll C$ when $\nu^* \ll 1$; otherwise, for a strong coupling $J_\perp \gg C$, only the regime of the second line takes place with a nearly constant $T_c(\nu)$. At $\nu > \nu^*$ the distance $1/\nu$ among solitons is shorter than the phase correlation length l_T, then the phase transition follows the Z_2 case scenario of pairwise binding of solitons; T_c generalizes the T_1 transition temperature. At $\nu < \nu^*$ (hence for any allowed $\nu < 1$ if $\nu^* > 1$) the solitons are rare over the phase correlation length l_T and the phase transition follows the XY regime of the phase ordering while isolated solitons are dressed, at $T < T_c$, by half-integer vortex rings.

Below T_c, there is a competition between the two forms of adaptation of solitons to the confinement. Because of the exchange of constituting particles, the chemical potentials are equilibrated. Then, the concentrations of bi-kinks and solitonic complexes are given as

$$\nu_{bk} = \exp(\beta(2\mu - 2E_s - 2NJ_\perp)) \, , \; \nu_s = \exp(\beta(\mu - E_s - E_V)) \, , \; 2\nu_{bk} + \nu_s = \nu ,$$

where $NJ_\perp$ is the interchain energy paid over the bi-kink, $E_V \propto T_{c0}$ is the energy of the vorticity accompanying the soliton. We can easily conclude that the total concentration is dominated by solitons, $\nu \approx \nu_s$, if: a) at any ν, T in the strong coupling case when $NJ_\perp \gg E_V$ and b) in the weak coupling case $NJ_\perp \ll E_V$ at sufficiently high $T > T^* = 2(E_V - NJ_\perp)/\ln(1/\nu)$. Neglecting the minor ν_{bk}, the condition $\mu - E_s = 0$ determines the phase transition of aggregation of solitons into the domain walls as $\nu = \exp(-E_V/T_2)$, $T_2 = E_V/\ln(1/\nu)$. This estimation works indeed for the strong coupling case. For the weak coupling, such a defined temperature T_2 is below the crossover, $T_2 < T^*$, where actually the concentration is dominated by the bi-kinks. Hence, with cooling below T_c, the ensemble of combined solitons evolves to the ensemble of bi-kinks while passing through the crossover T^*. Then, the bisolitonic mechanism of aggregation into double-wall rods takes place below the temperature estimated as $T_2' = NJ_\perp/\ln(1/\nu)$, analogously to the Z_2 case.

In the dimension $D = 2$, the above estimations of T^*, T_2 for the bi-kinks regime are valid except for the fact that T_2' becomes a crossover where the rods start to grow, rather than a sharp transition. Still, the solitonic mechanism of the strong coupling keeps the sharp transition. This nontrivial difference comes from long range pseudo-Coulomb interactions of half-integer vortices terminating a finite segment of aggregated solitons. We can explain that by extending simple arguments which have been proved for the case of the discrete symmetry by comparison with the exact solution available at $D = 2$ [6]. In the next section we shall illustrate them by means of the numerical modeling.

Consider a thermodynamically equilibrium system of transverse segments of solitonic walls of variable lengths l. Because of exchange of solitons among the rods, their chemical potentials are $\mu(l) = \mu l$. In $D = 2$, their energies $E(l)$ can be estimated as:

$$E(l) = E_s l + 2E_c + \tfrac{1}{2}\pi E_V \ln l. \qquad (8)$$

Here, E_s is the core energy of the amplitude soliton; $2E_c \sim J_\perp$ is the local energy of termination points and the last term is the pseudo-Coulomb interaction between terminating half-integer vortices. The distribution of walls over their lengths $n(l) = \exp((\mu(l) - E(l))/T)$ yields the total concentration of solitons as

$$\nu = \sum_{l=1}^{\infty} l\, n(l) = \exp\left(-\frac{2E_c}{T}\right) \sum_{l=1}^{\infty} l^{1-\alpha} e^{-ml} \, , \; m = \frac{(E_s - \mu)}{T} \geq 0 \, , \; \alpha = \frac{\pi E_V}{2T}. \qquad (9)$$

Recall the situation of the Z_2 case, where the expression (8) holds with $E_V = 0$, hence $\alpha = 0$ in (9). As $T \to 0$, the factor $\exp(-2E_c/T) \to 0$, so to keep ν fixed, the sum in (9) needs to diverge at $m \to 0$ which would happen at $\alpha < 2$. That holds indeed for the Z_2 case with $\alpha = 0$, then $\nu \propto \exp(-2E_c/T)/m^2$ which yields the characteristic rod length growing with T as

$$l \propto \frac{1}{m} \propto \nu \exp(E_c/T).$$

For the system with the vorticity when $E_V \neq 0$, the $m \to 0$ divergence requires that $T > \pi E_V/4 \simeq T_{c0}$. However, we are in the region well below T_{c0} where $\alpha > 2$, then the sum in (9) is bounded from above being convergent even at $m = 0$. Therefore, the ensemble of finite wall segments cannot further accommodate all the solitons, with the rest of them being stored in a reservoir played by domain walls crossing the sample. The temperature of this condensation is given by Equation (9) at $m = 0$ which yields $T_2 \approx 2E_c/\ln(1/\nu)$. Figure 4 shows the expected phase diagram of the system for the simpler case of the strong interchain coupling.

The $2D$ case is special because of a particular role of thermally activated vortices leading to a famous physics of the Berezinskii–Kosterlitz–Thouless (BKT) transition, see [19,20]. Here, our consideration is complicated by several factors. Phase fluctuations are essential even in the low T region, destroying the long range order while leaving the power law correlations. Vanishing of the order parameter at all temperatures leaves the mean field approximation for the upper transition at $T_c(\nu)$ (6) and (7) to be valid only locally. The thermally activated conventional integer vortices are abundant at high T, playing a leading role in the phase transition. These properties are well established for the pure XY model, which corresponds to our model in the limit $\nu = 0$.

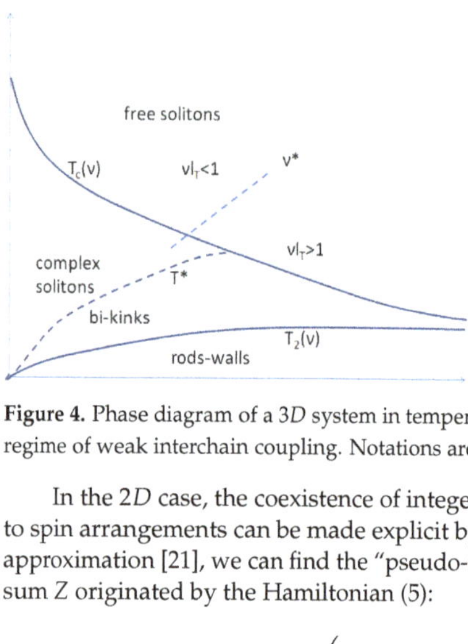

Figure 4. Phase diagram of a $3D$ system in temperature vs solitons' concentration coordinates in the regime of weak interchain coupling. Notations are explained in the text.

In the $2D$ case, the coexistence of integer and half-integer vortices and their relation to spin arrangements can be made explicit by the following procedure. Using the Villain approximation [21], we can find the "pseudo-Coulomb gas" representation of the statistical sum Z originated by the Hamiltonian (5):

$$Z = \sum_{\{m_R\}} \sum_{\{s_r\}} \exp\left(\pi\beta T_{c0} \sum_{R \neq R'} (m_R + \nu_R) \ln \frac{|\mathbf{R} - \mathbf{R}'|}{a} (m_{R'} + \nu_{R'}) \right.$$
$$\left. + (\pi^2 \beta T_{c0}/2) \sum_R (m_R + \nu_R)^2 + \beta J_{||} \sum_r s_r s_{r+\hat{x}} \right) \tag{10}$$

$$\nu_R = \frac{1}{4}(s_{r+\hat{x}} s_{r+\hat{x}+\hat{y}} - s_r s_{r+\hat{y}}) = \{0, \pm 1/2\} \tag{11}$$

Here, the indices r and R run over sites of the original lattice and the dual one correspondingly; the summation goes over all integer values of m_R and all configurations of spins $s_r = \pm 1$. The vortex "charge" (the vorticity) $m_R + \nu_R$ can acquire, unusually, a half-integer value; moreover, it becomes dependent on the local configuration of spins. Apart from the conventional long-range pseudo-Coulomb energy associated with all vortices, there is an additional local Ising term $\sim J_{||}$. The anomalous contribution to the vorticity ν_R, see Equation (11), at a site R emerges for particular configurations of spins s_r in the surrounding placket r: with three spins up and one down, or vice versa, see Figure 5. Such a configuration appears only as a termination of a transverse line of kinks: $s_r s_{r+\hat{x}} = -1$

(the orange lines in Figures 5a and 6), leading to the linearly growing, as $|y|J_{||}$, of the $\sim J_{||}$ term in (10). Thus, the longitudinal spin-coupling term with $J_{||} \neq 0$ leads to confinement of half-vortex pairs, situated at the ends of transverse solitonic walls, preventing formation of single half-vortices.

At the special value $J_{||} = 0$, the Hamiltonian (10) gives rise to a BKT type physics of half-vortices with the associated BKT transition of unbinding of their pairs at $T = T_{BKT}^{half} = T_{BKT}/4$. The point $J_{||} = 0$ can be realized, in principle, experimentally by coupling to an electronic bath with an appropriate tuning of the voltage difference. For a closed system, the limit $J_{||} \to 0$ is realized when the chemical potential of amplitude solitons is trapped at E_s, which happens below the temperature of the solitonic walls formation.

The opposite limit $\beta J_{||} \to \infty$ means that $(\mu - E_s)/T \to -\infty$, i.e., the vanishing concentration of solitons. In this limit, the half-vortices are strongly confined by the energy-costly spin strings. The game is dominated by conventional thermally activated integer vortices which drive the conventional BKT type transition at $T = T_{BKT} \approx 0.9C$ between the high-temperature disordered phase and the low-temperature phase with algebraically decaying correlations.

For intermediate $J_{||}$ we expect a crossover between regimes with domination of either integer or half-integer vortices.

The outlined phase diagram was confirmed via a numerical modeling by calculating macroscopic averages and their anomalies at phase transitions [12,13], and by displaying local pattern formations which examples we shall show below.

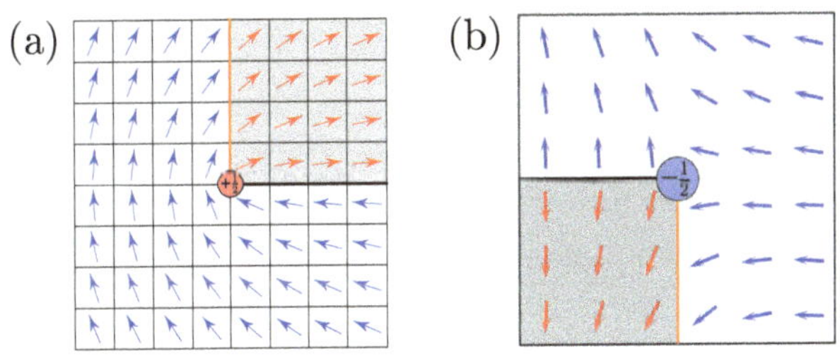

Figure 5. A half-vortex (**a**)—drawn schematically (after [13]), and a half-antivortex (**b**)-extracted from simulations of Figure 8. Here and in the following figures: arrows show XY degree of freedom, white/gray shading shows the Ising degree of freedom, the orange vertical (transverse to chains) thick line shows the array of amplitude solitons, the black horizontal (along the chains) thick line shows a string attached to a half-vortex—the line of transverse π-jumps in the phase θ, the disks denote the cores of half-integer (anti)vortices. The orange line is a uniquely defined physical object: the $J_{||}$ energy-costing spin mismatches. The black line is gauge dependent, lacking a physical singularity, with no energy cost.

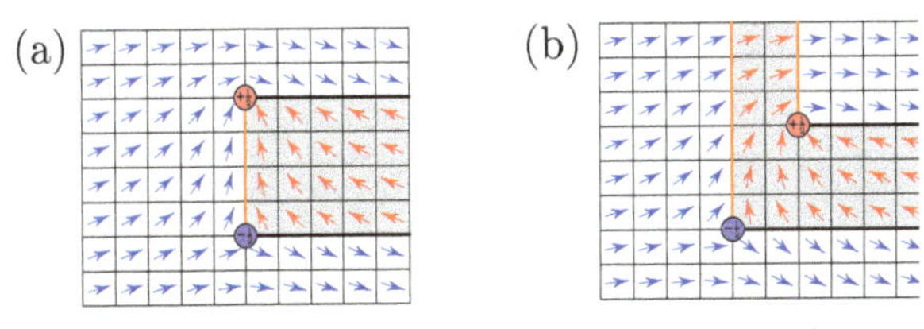

Figure 6. Half-integer vortex-antivortex pairs. (**a**) $T > T_2$, the pair structure is dominated by the linear confinement potential, solitonic walls (orange) are finite. (**b**) $T < T_2$, the pair structure is dominated by 2*D* Coulomb log-potential, solitonic walls grow to infinity. (After [13]).

3.3. *Pattern Formation from Numerical Simulations*

In this section, we describe the pattern formation in relation to the phase diagram. First, recall the terminology. An *amplitude soliton* (or just a *kink* for brevity) is an interruption of the Ising order along the chains' (x) direction, i.e., the soliton is present when $s_r s_{r+\hat{x}} = -1$. A *solitonic wall* (or a *line of kinks* in $D = 2$) is the aggregation of solitons transversely to the chains. (In the $3D$ case, the wall becomes a *solitonic disk*.) A *domain wall* appears when a solitonic wall grows across the whole transverse section of the sample. A *bisoliton* consists of two kinks at the minimal in-chain distance a between them; it is a reversed spin s_r in the otherwise ordered part of the chain. A *rod* in $2D$ is an aggregation of bisolitons arranged transversally to the chains. Analogous configurations in $3D$ are called "bisolitonic disks". When such discs grow transversially and reach the boundaries of the system, we obtain bisolitonic domain walls.

We model the cooling behavior of a system described by the Hamiltonian (5) using the Metropolis Monte Carlo method. The numerical algorithm preserves the number of amplitude solitons (so the term $J_{||} \sum_r s_r s_{r+\hat{x}} = const$) which substantially simplifies the analysis. There are no restrictions upon the phase degree of freedom θ_r.

Figure 7 shows a configuration with four solitons which are dressed by the phase adjustments $\theta \to \theta + \pi$, restoring the interchain order. Here, we observe two neutral half-VA pairs (outer ones) and also two charged half-VV and half-AA pairs (inner ones). The presence of such configurations with combined phase-amplitude topological defects grows with the increasing $J_\perp$. Sharper versions of such combined defects can be also found in a great number in Figure 8.

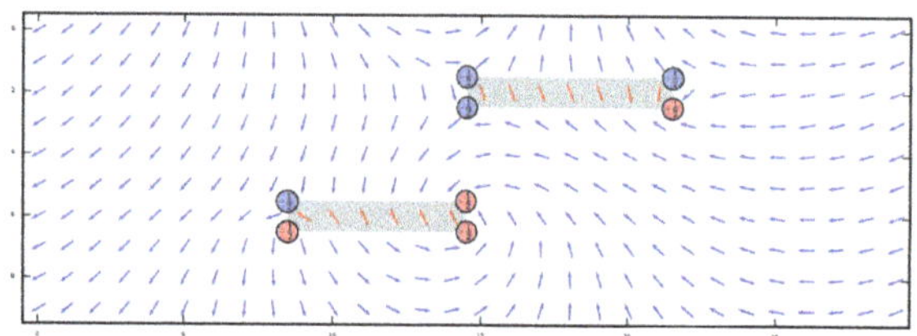

Figure 7. Configurations with four solitons for a system 10×30 with $J_\perp = 0.25C$, which was rapidly quenched down to $T \ll C$. Red and blue disks denote half-integer vortices and antivortices.

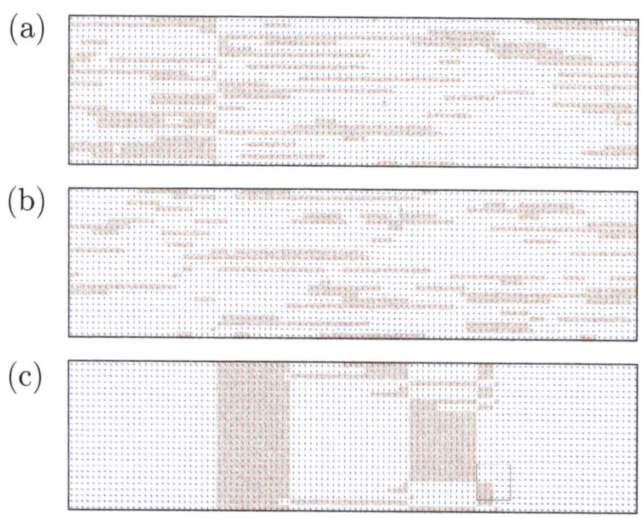

Figure 8. Configurations upon cooling for a system 25×100 with a strong interchain coupling $J_\perp = 4C$. (**a**) $T = 0.31C$, (**b**) $T = 0.30C$, and (**c**) $T = 0.01C$. From (**a**) to (**b**) we see that the transverse rod of solitons grows across the whole sample towards formation of macroscopic domains in (**c**). The box in (**c**) shows the half-antivortex, presented in Figure 5.

3.3.1. Quasi-Equilibrium Cooling

We begin with the strong interchain coupling: $J_{\perp} = 4C$. Cooling the system down starting from high temperatures, first we observe only short transverse solitonic walls (Figure 8a). At a temperature $T_2 \approx 0.3C$ we observe a sudden change: one of the walls starts to grow until it reaches the borders of the sample (Figure 8b). With a further cooling, more solitonic walls appear (Figure 8c) until the material for their building—free solitons—is exhausted. All the domain walls are installed through the growth of solitonic walls.

For a weak interchain coupling $J_{\perp} = 0.25C$, the walls formation is similar to the previously studied Z_2 case of the pure Ising model. The aggregation proceeds via proliferation of bisolitonic rods which do not need to be terminated by vortices (Figure 9).

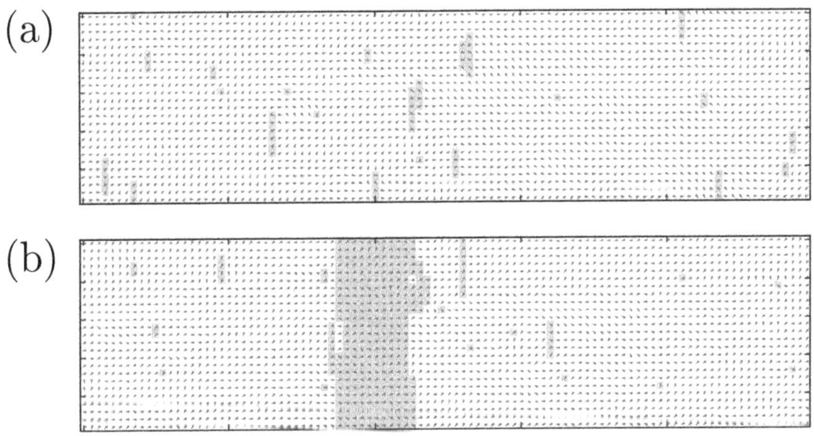

Figure 9. Configurations upon cooling for a weak interchain coupling $J_{\perp} = 0.25C$. (**a**) $T = 0.33C$, (**b**) $T = 0.32C$.

3.3.2. Quench versus Cooling: Two Regimes of Aggregation

In this section, we simulate a sudden quench from high to low temperatures $T < T_2$. This scenario is relevant to modern experiments on phase transitions induced by optical or electrical pulses, particularly in CDWs and other electronic crystals, see the discussion and references in [14]. We shall demonstrate that even for the case of weak interchain coupling the setting of domain walls is dominated by the mechanism of aggregation of soliton–vortex complexes.

Figure 10 shows configuration of a system with a weak interchain coupling $J_{\perp} = C/4$ after a sudden quench from a high temperature $T > T_2$ to the low temperature $T = 0.25C < T_2$. We observe the domain wall created by the solitonic mechanism, even if for the quasi-equilibrium slow cooling at the weak coupling the growth was dominated by the bisolitonic mechanism. When the solitonic wall initially starts to grow, half-vortices with opposite charges (vorticities) are created at its ends. These charges attract smaller solitonic rods via pseudo-Coulomb interaction mediated by the XY subsystem, promoting the further growth of the solitonic walls. The energy gain from merging of two walls is $\sim C \ln l$, so the growth process is self-accelerating and for finite samples the solitonic walls can even grow across the whole sample. On the contrary, the growth of bisolitonic rods goes via attaching of new bisolitons at the ends—similar to the case of the Z_2 symmetry, which is a much slower random walk process. At the same time, longer solitonic rods become low-mobile, therefore their recombination to bisolitonic rods is kinetically suppressed.

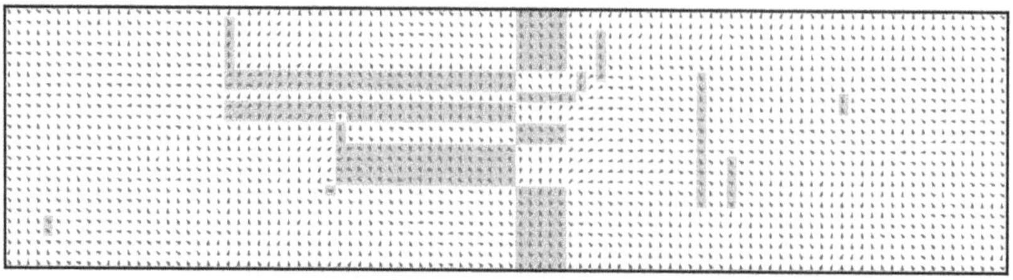

Figure 10. Configuration of the system after sudden quench from $T > T_2$ to $T = 0.25C < T_2$ (for a system 25×100 with $J_\perp = C/4$).

4. Conclusions

Solitons have returned to the agenda of studies of correlated electronic systems as their cooperative states can be created and studied in experiments on optical pumping and field effect doping. In this review, we have outlined results of the numerical and the qualitative analyses of phase transitions in ensembles of solitons. The numerical simulation of large-scale pattern formation in the full 3D system, including true Coulomb forces and for the combined discrete-continuous symmetry, was a challenging goal. The success came from using an efficient Monte Carlo algorithm which steps preserve the total number of solitons while allowing for their on-chain number variations. We have studied two of the most common cases:

I. The double degeneracy of the ground state which is found experimentally in electronic systems with a spontaneous dimerizations of bonds or charges. The solitons appear as kinks in the field of a scalar order parameter. In comparison with earlier studies [7], here we employed a more advanced numerical technique [12] which allowed us to model the pattern formation in $D = 3$, including the true Coulomb effects. Results of the modeling can be qualitatively summarized as following. The systems of solitons shows a sequence of phase transformations:

i. The confinement transition at T_1, below which individual solitons are bound into on-chain pairs. It is followed by progressive aggregation of bisolitons into transverse disk-like or rod-like formations;
ii. At a lower critical temperature T_2, the second phase transition (a crossover in $D = 2$) occurs: the bisolitonic disks cross the entire sample and disintegrate into solitonic domain walls;
iii. For electrically charged solitons, even locally small Coulomb interactions can nevertheless affect the T_2 transition where macroscopic patterns are created. A large scale structure, such as a domain wall, gives rise to a high long-range electric field withstanding the confinement forces. That erases the gain of the confinement energy reached by the wall formation, resulting in fragmenting macroscopic patterns which give rise to long-range electric field.

II. The combined degeneracy with respect to both the amplitude and the phase of the order parameter. The motivation of presented studies was to notice that most common types of symmetry breaking (superconductivity, spin density waves, antiferromagnetic Mott state, incommensurate charge density waves) in quasi-one-dimensional electronic systems possess a combined manifold of degenerate states. Beyond the standard continuous XY-type degeneracy with respect to the phase θ degree of freedom, there is also an Ising-type discrete degeneracy with respect to the sign of the amplitude A of the order parameter $A\exp(i\theta)$. These two degrees of freedom can be controlled or accessed independently via either the spin polarization or the charge doping. The degeneracies give rise to two coexisting types of topologically nontrivial configurations: phase vortices or amplitude kinks – the solitons. Being decoupled at the 1D level of isolated chains, at higher D the two degrees of freedom experience confinement which binds together kinks and half-integer vortices. These combined amplitude-phase solitons are the lowest-energy excitations of either spin or charge taking that role from conventional electrons. The case of the combined symmetry also shows a sequence of two phase transitions. At the higher T_1 the (quasi in $D = 2$) long-range order of the XY type sets in, leading to the confinement which dresses the kinks by half-integer vortices. At the lower T_2 the liquid of combined solitons starts to structure by aggregating the solitons into walls growing across the sample. That proceeds via either solitonic domain walls or bisolitonic rods. The growing solitonic walls are formed by transversely correlated amplitude kinks; termination points of walls of finite lengths must be accomplished by half-integer vortices. Attractions of vortices from different walls may force the walls to glue together forming a topologically trivial bisolitonic rod. For non-equilibrium processes, such as supercooling by a sudden quench, the solitonic-wall scenario can be strongly promoted because the growing solitonic wall has uncompensated half-vortex "charges" at the ends. They attract mobile smaller rods, while the slow process of gluing of low-mobile long solitonic walls into bisolitonic rods can be kinetically suppressed.

Author Contributions: All authors contributed equally to the presenting research and preparing the publication. All authors have read and agreed to the published version of the manuscript.

Funding: This research received no external funding.

Institutional Review Board Statement: Not applicable

Informed Consent Statement: Not applicable

Data Availability Statement: Not applicable

Acknowledgments: P.K. acknowledges the support by Max Planck Society.

Conflicts of Interest: The authors declare no conflict of interest.

References

1. Manton, N.; Sutcliffe, P. *Topological Solitons*; Cambridge University Press: Cambridge, UK, 2004.
2. Brazovskii, S. New Routes to Solitons in Quasi One-Dimensional Conductors. *Solid State Sci.* **2008**, *10*, 1786–1789. [CrossRef]
3. Brazovskii, S. Microscopic solitons in correlated electronic systems: Theory versus experiment. Advances in Theoretical Physics: Landau Memorial Conference. *AIP Conf. Proc.* **2009**, *1134*, 74.
4. Mineev, V.P. *Topologically Stable Dedects and Solitons in Ordered Media*; Harwood Academic Publishers: Harwood, MD, USA, 1998.
5. Mermin, N.D. The topological theory of defects in ordered media. *Rev. Mod. Phys.* **1979**, *51*, 591. [CrossRef]
6. Bohr, T., Brazovskii, S. Soliton statistics for a system of weakly bound chains: Mapping to the Ising model. *J. Phys. C Solid State Phys.* **1983**, *16*, 1189. [CrossRef]
7. Teber, S.; Stojkovic, B.P.; Brazovskii, S.; Bishop, A.R. Statistics of charged solitons and formation of stripes. *J. Phys. Condens. Matter* **2001**, *13*, 4015. [CrossRef]
8. Teber, S. Statistical properties of charged interfaces. *J. Phys. Condens. Matter.* **2002**, *14*, 7811. [CrossRef]
9. Korshunov, S.E. Possible splitting of a phase transition in a 2D XY model. *JETP Lett.* **1985**, *41*, 263.
10. Korshunov, S.E. Phase diagram of the modified XY model. *J. Phys. C Solid State Phys.* **1986**, *19*, 4427. [CrossRef]
11. Grudst, F.; Pollet, L. Z_2 parton phases in the mixed-dimensional $t - J_z$ model. *Phys. Rev. Lett.* **2020**, *125*, 256401.
12. Karpov, P.; Brazovskii, S. Phase transitions in ensembles of solitons induced by an optical pumping or a strong electric field. *Phys. Rev. B* **2016**, *94*, 125108. [CrossRef]
13. Karpov, P.; Brazovskii, S. Phase transitions and pattern formation in ensembles of phase-amplitude solitons in quasi-one-dimensional electronic systems. *Phys. Rev. E* **2019**, *99*, 022114. [CrossRef] [PubMed]
14. Karpov, P.; Brazovskii, S. Modeling of networks and globules of charged domain walls observed in pump and pulse induced states. *NPJ Sci. Rep.* **2018**, *8*, 4043. [CrossRef] [PubMed]
15. Christov, I.C.; Decker, R.J.; Demirkaya, A.; Gani, V.A.; Kevrekidis, P.G.; Radomskiy, R.V. Long-range interactions of kinks. *Phys. Rev. D* **2019**, *99*, 016010 [CrossRef]
16. Kwon, H.-J.; Yakovenko, V.M. Spontaneous formation of a π soliton in a superconducting wire with an odd number of electrons. *Phys. Rev. Lett.* **2002**, *89*, 017002. [CrossRef] [PubMed]
17. Granto, E.; Kosterlitz, J.M.; Lee J.; Nightingale M.P. Phase transitions in coupled XY-Ising systems. *Phys. Rev. B* **1991**, *66*, 1090.
18. Lee J.; Granto, E.; Kosterlitz, J.M. Nonuniversal critical behavior and first-order transitions in a coupled XY-Ising model. *Phys. Rev. B* **1991**, *44*, 4819. [CrossRef] [PubMed]
19. Jose, J.V. *40 Years of Berezinskii–Kosterlitz–Thouless Theory*; World Scientific: Singapore, 2013.
20. Kosterlitz, J.M. Kosterlitz–Thouless physics: A review of key issues. *Rep. Prog. Phys.* **2016**, *79*, 026001. [CrossRef] [PubMed]
21. Villain, J. Theory of one- and two-dimensional magnets with an easy magnetization plane. *J. Phys.* **1975**, *36*, 581. [CrossRef]

symmetry

MDPI

Article

Simulations of Dynamical Electronic Vortices in Charge and Spin Density Waves

Natasha Kirova [1,2,*] and Serguei Brazovskii [3]

1 CNRS, LPS, University Paris-Saclay, 91405 Orsay, France
2 Russian Quantum Center, Skolkovo, Moscow 143025, Russia
3 CNRS, LPTMS, University Paris-Saclay, 91405 Orsay, France
* Correspondence: natacha.kirova@universite-paris-saclay.fr

Abstract: Charge and spin density waves are typical symmetry broken states of quasi one-dimensional electronic systems. They demonstrate such common features of all incommensurate electronic crystals as a spectacular non-linear conduction by means of the collective sliding and susceptibility to the electric field. These phenomena ultimately require for emergence of static and transient topological defects: there are dislocations as space vortices and space-time vortices known as phase slip centers, i.e., a kind of instantons. Dislocations are statically built-in under a transverse electric field; their sweeping provides a conversion among the normal carriers and condensate which ensures the onset of the collective sliding. A special realization in a high magnetic field, when the density wave is driven by the Hall voltage, originated by quantized normal carriers, reveals the dynamic vorticity serving to annihilate compensating normal and collective currents. Spin density waves, with their rich multiplicative order parameter, bring to life complex objects with half-integer topologically bound vorticities in charge and spin degrees of freedom. We present the basic concepts and modelling results of the stationary states and their transient dynamics involving vorticity. The models take into account multiple fields in their mutual non-linear interactions: the complex order parameter, the self-consistent electric field, and the reaction of normal carriers. We explore the traditional time-dependent Ginzburg–Landau approach and introduce its generalization allowing the treatment of intrinsic normal carriers. The main insights and illustrations come from numerical solutions to partial differential equations for the dissipative dynamics of one and two space dimensions.

Keywords: charge density wave; spin density wave; electronic crystal; topological defect; vortex; dislocation; phase slip

1. Introduction

Embedded or transient topologically non-trivial configurations [1,2] are common among symmetry broken ground states [3]. Numerous applications involve particle physics [4], cosmology [5,6], quantum liquids [7–9], gases [10], optical condensates [11], cold atoms [12], and liquid crystals [13,14].

More specifically, solid states bring to life domain walls and discommensurations in superstructures [15], current vortices in superconductors [16], displacements vortices as dislocations [17], phase slips and vortices in sliding superstructures (discussed in this article), and walls and skyrmions in magnetic media [18–20]. The ground states with spontaneously broken symmetries are ubiquitous in correlated electronic systems. Among them is the vast family of electronic crystals (see review [21] or collection [22]) which includes charge and spin density waves (CDW, SDW), Wigner crystals, stripe arrays, charge ordering states, electronic ferro- and antiferroelectrics, and superstructures in spin singlet systems under a magnetic field: spin-Peierls state, spin-polarized density waves (DW) and superconductors. The ground state degeneracy tolerates the formation of various topologically protected configurations connecting equivalent, but different, ground states. These topological defects include macroscopic extended objects (plane domain walls, dislocation lines, phase vortices), and microscopic ones (various solitons and instantons.)

The electronic crystals can be incommensurate with the underlying host lattice; consequently, their number of unit cells is not fixed. This allows them to absorb excess electrons into the extended (with a different number of total DW periods) ground state. These superlattices, particularly incommensurate CDWs, SDWs [21,23–25] and Wigner crystals [21], are able to perform the collective current conduction (Frölich conductivity [21,23,26]) by means of sliding. Here, the topological defects are crucial for the conversion between normal and collective currents and they also appear in de-pinning processes to initiate sliding in the presence of host defects and constraints. The exchange among normal and condensed charge carriers comes from their imbalance or a stress from the electric field. The steps involve topologically non-trivial objects, such as amplitude and phase solitons, and phase vortex dislocations, which proceed via transient phase slip processes, i.e., the space-time vortices [27,28]. This results in a rich complex of non-linear and non stationary behaviours with vast experimental observations.

In this article, we mostly address DWs which are ubiquitous in quasi one-dimensional (1D) systems [23,24]. We concentrate on a phenomenological description, which is already quite complicated involving coupled equations for order parameters, the self-consistent electric potential, and the density and current of normal carriers interacting with both fields. The main tool used are numerical solutions to these equations, yielding configurations involving static and dynamic topological defects.

The structure of the article is as follows. In Section 2, we outline the basic properties and regimes of DW sliding and deformation. In Section 2.1, we describe the approach based on intuitive phenomenological equations for the CDW order parameter interacting with the self-consistent electric field and extrinsic normal carriers. In the Section 3 we introduce the generalized GL phenomenology. It takes into account explicitly the effects of intrinsic normal carriers and avoids problems with the conservation of condensed density, inherent to the traditional approach of the Section 2.1. Section 4 addresses the special circumstance of CDW sliding driven by the Hall voltage under a high magnetic field. Section 5 discusses more complicated configurations allowed for the rich order parameters of the SDW. We address the scenario of splitting conventional dislocations into combined objects of topologically bound half-integer vortices of the phase and magnetic components. Finally, Section 6 summarises our conclusions.

2. Basic Properties and Regimes in Sliding Density Waves

CDWs [23,24] and SDWs [25] are directly observed as sinusoidal periodic modulations which perturb the electronic density together with atomic distortions for CDWs or electronic spin polarization $\vec{m}$ for SDWs. The DW deformations are characterized by the superstructure wave vector $\vec{Q}$, the normalized amplitude $A(x,t)$, and the phase $\phi(x,t)$. These macroscopic quantum states [26] can be described by a complex scalar-order parameter for the CDW and a complex vector for the SDW:

$$\eta_{CDW}(\vec{R},\phi) = A\exp(\vec{Q}\cdot\vec{R}+\phi)\ ;\ \eta_{SDW} = A\vec{m}\exp(\vec{Q}\cdot\vec{R}+\phi). \tag{1}$$

Phenomenologically, a quasi 1D DW can be viewed as an elastic uniaxial crystal with distortions of the phase $\phi(\vec{R},t)$ playing the role of conventional displacements. The relationships between DWs and the traditional notion of conventional crystals, their elasticity and topological defects, such as dislocations [29,30], are summarized in Table 1.

Table 1. Relationships between DWs and conventional crystals.

Conventional crystal	DW (CDW or SDW)
Atomic density ρ_{atomic}	DW modulation: $A\cos(\vec{Q}\cdot\vec{R}+\phi)\times(1 \text{ or } \vec{m})$
Displacements $\vec{u}/a$	$-\vec{v}\phi/2\pi$, $\vec{v}=(1,0,0)$
Compression $-\nabla\vec{u}$	DW charge density $\frac{1}{\pi}\frac{\partial\phi}{\partial x}$
Velocity $\frac{\partial\vec{u}}{\partial t}$	DW current $-\vec{v}\frac{1}{\pi}\frac{\partial\phi}{\partial t}$
Vacancies or addatoms	$\pm 2\pi$ solitons
Dislocations	Phase vortices (plus $\vec{m}$ vortices in SDWs)

Here a is the host crystal period, $\vec{Q}$ is the DW equilibrium wave number, and $\vec{m}$ is the unit vector of the staggered magnetization in a SDW, where $\vec{R}=(x,\vec{r}_{\perp})$ (with x being the coordinate in the chains' direction $\vec{v}$ and location $\vec{r}_{\perp}$).

As DWs are electronic crystals, their deformations are charged giving rise to long-range Coulomb forces which result in anomalous elasticity. Properties of all macro- and microscopic topological defects, their interactions, and transformations, are strongly affected by the Coulomb energy [31,32].

Commonly, the sliding DW current is driven by the electric field E_x along the chains; therefore, the collective and normal currents flow in the same direction x. The current conversion near contacts progressively gives rise to the collective current, resulting in a stationary equilibrium in the bulk. The conversion of the injected current of normal electrons (or holes) into the collective DW current and back requires adding electronic crystal planes near the contact source and their elimination near the drain. In thick channels that proceeds via a transverse flow of dislocations [33,34] which can be viewed as the leading edges of the penetrating/retreating planes, while in wires the plane phase slips are expected to be formed (see [28] for a review). In any case, an extended distribution of these topological events is formed, as identified in CDWs by space-resolved X-ray diffraction [35,36].

Dislocations, as electronic vortices, can be initiated statically when the lateral electric field $E_{y,z}$ is applied transversely to the chains' orientation [37], or due to variations in $\vec{Q}$ near the surface [38], or because of thermal hysteresis of $\vec{Q}$ [39]. All of this provides a rich complex of non-linear and non stationary effects, partly illustrated in Sections 2.1 and 3.

In superfluids or superconductors, vortices can move in any direction transverse to their axis. However, for dislocations there are two types of motion [17]: conservative (glide) and non-conservative (climb). A glide along the Burgers vector passes freely, while a transverse to the Burgers vector climb requires an inflow of addatoms or vacancies (electrons and holes, or their polaronic states [15] for electronic crystals). Specifically for DWs, when a dislocation climbs across a chain, the phase increment along the chain changes by $\pm 2\pi$, i.e., one period is added/removed; thus, two electrons are absorbed/emitted from/to the reservoir of normal carriers. Thus, the climb of dislocations, the rarest process in conventional crystals, becomes obligatory in sliding electronic crystals, while the common glide is less important. Considering these events on average, we arrive at a new multi-fluid hydrodynamic system involving conventional and topological objects [40].

A peculiar realization [41] of CDW sliding can be achieved using mesa-junctions under an external high magnetic field, where sliding can be driven by the Hall voltage V_H originating from the transverse current of normal carriers. Here, unusually, the collective and normal currents move in opposite directions, allowed to be exactly compensated for so that the total current is zero. The zero-sum loop of the collective current is compensated exactly by the counter current of normal electrons. This promotes their annihilation at the boundaries of the open circuit design, which is provided by periodic phase slip processes seen as spontaneous coherent oscillations.

The rich order parameter of the SDW allows for complex topological defects where phase displacements interfere with the magnetic orientation [42]. Thus, the half-integer π-vortices or phase slips can appear in SDWs, while they are forbidden in CDWs where only self-mapping 2π phase circulations are allowed. With assistance from Coulomb repulsion forces, the conventional dislocations lose their priority bringing to life an unusual topological object, constructed by the combination of a half-integer dislocation and a semi-vortex of the staggered magnetization.

DW deformations and movements are realized through interacting and exchanging with a sea of normal carriers present as extrinsic n_{ex} and intrinsic n_{in} carriers. The extrinsic carriers come from pockets on the Fermi surface which are not connected by the CDW wave number $\vec{Q}$ or even belong to another electronic band; therefore, they do not participate in DW formation. They only interact with the charged DW deformations via the common electric potential Φ. The intrinsic carriers belong to bands participating in DW formation; they can be activated across the DW gap or injected, thus mutually influencing the DW amplitude. Importantly in our context, they are also affected by DW phase deformations, so that the total potential energy V for intrinsic carriers can be written as:

$$V = \Phi + \delta E_F \, , \; \delta E_F = v_F \delta p_F = \frac{1}{\pi N_F} \frac{\partial \phi}{\partial x} = \frac{n_c}{N_F} \, , \; N_F = \frac{2}{\pi \hbar v_F} \tag{2}$$

where v_F and N_F are the Fermi velocity and the related DOS in the parent metal, respectively, and δE_F is the Fermi energy shift coming from the DW briefing $\sim \partial_x \phi$.

With this in mind, our numerical simulations are presented in two approaches:

(i) the minimal Ginzburg-Landay type model, where all intrinsic carriers are prematurely integrated out [43,44];
(ii) a deeper and more complicated model [45] exposing the intrinsic carriers explicitly.

In both cases, we take into account multiple fields in their mutual non-linear interactions: the amplitude A and the phase ϕ of the order parameter, distributions of the electric potential Φ, of the density n, and the current j of normal carriers. The results of the numerical modelling for the static, stationary, and transitory effects in CDWs are presented in Sections 2.1 and 3.

The approaches tested for CDWs are extended in Section 5 to a more complicated SDW case. For both CDW and SDW, a spontaneous origin of dislocations and vortices is achieved without seeding or promoting their formation.

2.1. The Ginzburg–Landau Type Model for the CDW

In this section, we study the CDW dynamics within the dissipative time dependent Ginzburg–Landay (TDGL) type model. Generically this approach can be traced back to the phenomenological equation $\gamma \partial_t \eta = -\delta W / \delta \eta$ for relaxation of the order parameter η determining the energy functional $W\{\eta\}$. It has been widely used (analytically, e.g., [46] and numerically, e.g., [10,47]) for non-equilibrium pattern formations, particularly with the appearance of topological defects when quenching across a phase transition. The parameter γ can be complex, interpolating with the Gross–Pitaevskii equation for a Bose condensate, allowing the inclusion of the quantum coherent states of polaritons or cold atoms. In this approach for fermionic systems, originally developed for superconductivity (see [16,27]) and extended to CDWs (see [48,49]), the intrinsic carriers are integrated out giving rise to coefficients of amplitude expansion and gradients of the order parameter; while extrinsic carriers from the bands not involved in CDW formation are taken into account explicitly (if present).

The density (per unit length of one chain) w_{GL} of the energy functional can be written as:

$$\begin{aligned} w_{GL} \; &= \tfrac{\hbar v_F}{4\pi} \{ (\partial_x A)^2 + \beta_y (\partial_y A)^2 + \beta_z (\partial_z A)^2 \\ &+ A^2 (\partial_x \phi)^2 + \beta_y A^2 (\partial_y \phi)^2 + \beta_z A^2 (\partial_z \phi)^2 \} + \\ &F(A) + \tfrac{\Phi}{\pi} A^2 \partial_x \phi + \Phi n_{ex} + F_{ex}(n_{ex}) - \tfrac{\varepsilon_{hst} s_\perp}{8\pi e^2} (\nabla \Phi)^2 \end{aligned} \tag{3}$$

Here the overall coefficient $\hbar v_F / 4\pi$ comes from the microscopically derivable terms $A^2 (\partial_x \phi)^2$ and $A^2 \partial_x \phi \Phi / \pi$, parameters $\beta_{y,z}$ describe CDW interchain coupling, $F_{ex}(n_{ex})$ is the free energy of extrinsic carriers, and ε_{host} is the host dielectric constant. The coupling of the phase with the electric potential Φ comes from the fact that within the Ginzburg–

Landau scheme $A^2\partial_x\phi e/\pi = n_c$ is the charge density distribution originated from phase deformations. The amplitude A is normalized such that the energy of a homogeneous state $F(A)$ has a minimum at $A = 1$. Various forms for $F(A)$, such as generic GL expansion F_{GL} or the broader energy dependence F_{PF} of the Peierls–Froehlich microscopic theory, can be input akin to the BCS:

$$F(A)\ : F_{GL} = \frac{\Delta_0}{2\xi_0}(-aA^2 + bA^4) \text{ or } F_{PF} = \frac{\Delta_0}{2\xi_0}A^2\ln\frac{A^2}{e}. \tag{4}$$

where Δ_0 and $\xi_0 = \hbar v_F/\Delta_0$ are the microscopic parameters: the electronic gap and correlation length, and a, b are numerical coefficients (functions of T).

The time evolution, assuming a dissipative regime, is described by functional derivatives of W in Equation (3) with respect to A, ϕ and Φ:

$$\frac{\hbar v_F}{4\pi}\left[(\partial_x^2 A + \beta_y\partial_y^2 A) - A(\partial_x\phi)^2 - \beta_y A(\partial_y\phi)^2\right] - \partial_A F(A) = \gamma_A\partial_t A \tag{5}$$

here and below, for brevity we only show derivatives over one transverse coordinate y.

$$\partial_x(A^2\partial_x\phi + \pi N_F\Phi) + \beta_y\partial_y(A^2\partial_y\phi) = -\gamma_\phi A^2\partial_t\phi \tag{6}$$

$$-\nabla^2\Phi = \frac{1}{N_F r_0^2}\left(\frac{A^2}{\pi}\partial_x\phi + n_{ex}\right) \tag{7}$$

The concentration of normal carriers obeys the diffusion equation

$$\nabla j_{ex} + \partial_t n_{ex} = -\nabla[\sigma_{ex}\nabla(\zeta_{ex} + \Phi)] + \partial_t n_{ex} = 0\,,\ \zeta_{ex} = \frac{\partial F_{ex}}{\partial n_{ex}} \tag{8}$$

Notice that, contrary to common minimal Ginzburg–Landau models [16,27,49] that only contain equations such as (5) and (6) for the order parameter, we include two more: the Poisson Equation (7), coupling collective and normal extrinsic charges through the potential Φ; and the diffusion equation for the carriers (8). Boundary conditions for these partial differential equations (see [43,44]) take into account that the normal CDW stress and normal electric field (confined within samples) vanish at all boundaries, and the normal current flow only exists at source–drain boundaries.

Below we discuss the results of the numerical solutions to the above equations in a rectangular geometry for two cases corresponding to different applications of the electric voltage: longitudinal and transverse.

For a transversely applied electric field, our simulations (see [43,44] for more examples) recover a fast initial creation of multiple dynamical vortices evolving to their smaller equilibrium number. The snapshots of a multi-vortex transient process in Figure 1 (left panel) illustrate vortices nucleations at opposite boundaries and their traversing of the sample with their mutual annihilation or absorption at another boundary (plot *a*). Only a small number of static vortices are left at the final stage (plot *b*). An isolated vortex is shown in detail in the right panel.

For a longitudinally applied voltage, the main interest lies in observing the space-time vorticity giving rise to phase slip processes. This one-dimensional regime has already been addressed in numerical simulations [50–52] based on minimalistic equations derived microscopically [28,49] for a dirty limit near the transition temperature. Contrarily, our multiple equations were designed for a pure system with a well-established gap in the fermionic spectrum where the self-consistent electric field and the reaction of normal carriers become important. For shortness, in Section 4 we demonstrate results for a special case of sliding under an HMF.

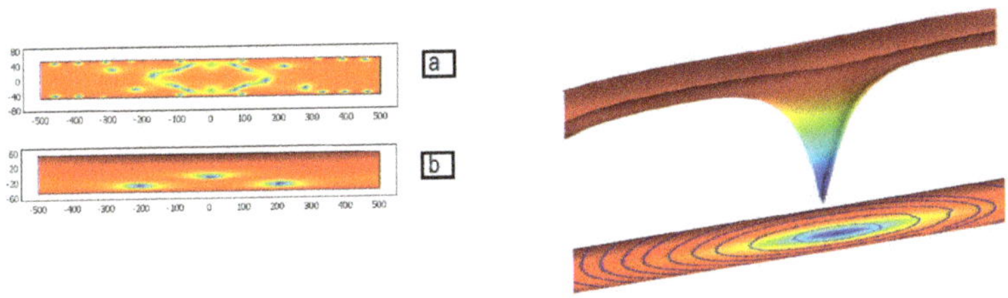

Figure 1. Left panel: 2D presentation of the CDW amplitude A. Upper plot (**a**): a snapshot of the transient state with many intermittent vortices. Lower plot (**b**): the final state with only three remaining vortices showing a tendency to form a triangular lattice. Right panel: 3D plot of the amplitude of a single CDW vortex in the final state. On the basis: the density plot of A with stream lines of the phase.

2.2. Coulomb Hardening, Anomalous Energy and Shape of Charged Dislocations in DWs

The strong contribution of Coulomb interactions to the energetics of dislocations is very important for their properties which may have a qualitative effect as will be discussed in Section 5. Here, we shall provide some necessary estimations which can be tuned by analytical solutions to some cases [31,37].

In a static linear regime with a constant amplitude (i.e., sufficiently far away from the amplitude node), the Coulomb potential and the normal carriers can be excluded from the GL free energy (3). In the Fourier phase representation ϕ_k, the resulting energy can be written as

$$W_{chrg}\{\phi\} = \frac{\hbar v_F}{4\pi a_\perp^2} \sum_{\vec{k}} |\phi_{\vec{k}}|^2 \left[C_\parallel^0 k_\parallel^2 + C_\perp k_\perp^2 + \frac{r_0^{-2} k_\parallel^2}{k_\parallel^2 + k_\perp^2 + r_{scr}^{-2}} \right] \tag{9}$$

with

$$C_\perp \sim \left(\frac{u_\perp T_c}{\hbar v_F}\right)^2, \; r_{scr} = \frac{r_0}{\rho_n^{1/2}}, \; r_0 = \frac{a_\perp}{(8\alpha)^{1/2}}, \; \alpha = \frac{e^2}{\hbar v_F}$$

where $C_\parallel^0 \sim 1$ is the compression modulus, $C_\perp$ is the share modulus which comes from the interchain coupling yielding the 3D transition temperature T_c. The last term is due to Coulomb interactions of charged phase deformations, taking into account the screening by normal carriers.

Let the phase be deformed in a volume with characteristic lengths $L_\parallel$ and $L_\perp$ determined, for example, by a dislocation loop or a dipole of dislocation lines. At distances below the screening length $L_\perp < r_{scr}$ the energy (9) can be estimated as

$$W_{dsl} \approx \frac{\hbar v_F}{4\pi a_\perp^2} \left[\frac{C_\perp}{L_\perp^2} + \frac{L_\perp^2}{r_0^2 L_\parallel^2} \right] L_\perp^2 L_\parallel \, , \; L_\perp < r_{scr}.$$

At the monitored $L_\perp$, the dependence of W_{dsl} on $L_\parallel$ is non-monotonous. Minimization over $L_\parallel$ results in the optimal length and energy

$$L_\parallel = C_\perp^{-1/2} L_\perp^2 r_0^{-1} \sim T_c \alpha^{1/2} (L_\perp / a_\perp)^2 \; , \; W_{dsl}\{L_\perp < r_{scr}\} = E_0 N$$

where $E_0 \sim \alpha^{1/2} T_c$ and $N \sim L_\perp^2 / a_\perp^2$ are the number of chains embraced by the dislocation loop. We can see that for non-screened Coulomb interactions the dislocation energy obeys the area law ($\propto N$), growing much faster than according to the usual perimetric law ($\sim N^{1/2} \ln N$) law.

At large distances $r_\perp >> r_{scr}$, the analytical (in $\vec{k}$) form of the elastic energy is restored, while with an enhanced compression module:

$$W_{chrg}\{\phi\} \approx \frac{\hbar v_F}{4\pi a_\perp^2}\sum_{\vec{k}}|\phi_{\vec{k}}|^2\left[C_\perp k_\perp^2 + \frac{k_\parallel^2}{\rho_n}\right]$$

At low T the normal carriers are frozen out, $\rho_n \to 0$, causing the divergence of the comprehensibility $C_\parallel = 1/\rho_n$ ("Coulomb hardening" effect [53]) under the simultaneous action of Coulomb interactions and screening carriers. For a dislocation, the standard perimetric law is restored $W_{dsl}\{L_\perp >> r_{scr}\} = E_0\rho_n^{-1/2}N^{1/2}\ln N$ with the dislocation energy enhanced as $\rho_n^{-1/2}$. Nevertheless, the creation of dislocations, in spite of their big energy at a small ρ_n, is allowed requiring higher applied potentials.

For another important geometry, consider a straight dislocation line positioned at a depth Y below the lateral junction surface. For screened Coulomb interactions at $Y \gg r_{scr}$, the dislocation energy (per unit of its length) $W_{disl} \sim E_0/a_\perp(r_{scr}/r_0)\ln(Y/r_{scr})L_x$ obeys the traditional vortex lines logarithmic law, but quantitatively it is greatly enhanced as r_{scr}/r_0. At low T there is an intermediate wide region $r_0 < Y < r_{scr}$ where a confinement law is set up with $W_{disl} \sim E_0 Y/a_\perp^2$. A constant force with a linear density $F_{conf} \sim E_0 Y/a_\perp^2$ acts upon the dislocation line requiring the application of an electric field with a definite threshold.

3. Generalized Ginzburg–Landau Type Model for CDW Vortex Formation

Within the traditional GL approach presented in the previous section, the intrinsic electrons are excluded ("fermions are integrated out") at the microscopic level. All observable quantities characterizing the CDW are determined via the order parameter alone. Now we consider the generalized GL-type theory [42,45] which explicitly accounts for intrinsic carriers and their interaction with the order parameter. The theory can be further extended [40] to include the possible extraction/reabsorption of normal carriers from/to the CDW collective density which proceeds via the formation of topological defects. This approach provides deeper insights into the partitions of the normal and collective charges and currents at the expense of specific numerical modelling difficulties as the theory is now non-analytic with respect to the order parameter.

The on-chain energy density $\tilde{w}_{GL}$ can be written as [42,45]:

$$\tilde{w}_{GL} = \frac{1}{2N_F}\left[(\partial_x A)^2 + \left(\frac{1}{\pi}\partial_x\phi\right)^2\right] + \frac{1}{\pi}\Phi\partial_x\phi + (\Phi + \frac{1}{\pi N_F}\partial_x\phi)n - \frac{\varepsilon_{host}s}{8\pi}(\nabla\Phi)^2 + F(A,n) \tag{10}$$

Here, for brevity we do not show terms related to transverse gradients or extrinsic carriers which are the same as in (3); also we omit the index "in" as $n_{in} \longrightarrow n$. The order parameter amplitude A is connected with n via the energy $F(A,n)$ which must have a property that its minimum over A disappears when n surpasses a critical value n_{cr} above which the CDW state is erased. Unlike the generic GL case, the terms containing $\partial_x\phi$ and $(\partial_x\phi)^2$ are not multiplied by A^2 making them singular near the vortex core when $A \to 0$. Originating from the chiral anomaly [42,45], they cannot be derived perturbatively. However, altogether the contribution of non-analytic terms vanishes when $A \to 0$. This hidden compensation can be proven by excluding n from the equations of motion which, in general, cannot be performed explicitly, rather than with the help of additional approximations [45].

The time evolution equations acquire the form (we only show only one transverse coordinate y):

$$\partial_x(\partial_x\phi + n + \pi N_F\Phi) + \beta_y\partial_y(A^2\partial_y\phi) = -\gamma_\phi A^2\partial_t\phi \tag{11}$$

$$-(\partial_x^2 A + \beta_y\partial_y^2 A) + \beta_y A(\partial_y\phi)^2 + \partial_A F(A,n) = -\gamma_A\partial_t A \tag{12}$$

$$\nabla^2\Phi = -\frac{1}{N_F r_0^2}(\frac{1}{\pi}\partial_x\phi + n) \tag{13}$$

Contrary to the former GL equations described in the previous section, now Equations (11)–(13) are non-analytic concerning the order parameter [45].

As before, we consider the kinetics of the carriers in the quasi-equilibrium diffusive approximation with the important peculiarity that they experience combining the potential V from (2), rather than just Φ:

$$\nabla j + \partial_t n = -\nabla(\hat{\sigma}\nabla\mu) + \partial_t n = 0\,,\ \mu = \zeta + \Phi + \frac{1}{\pi N_F}\partial_x\phi\,,\ \zeta = \frac{\partial F(A,n)}{\partial n} \tag{14}$$

Here $\hat{\sigma}$, ζ and μ are the conductivities, the chemical potential, and the electrochemical potentials of intrinsic carriers, respectively. We define the normal ρ_n and collective ρ_c charge densities as

$$\rho_n = \frac{1}{N_F}\frac{\partial n}{\partial \zeta}\,,\ \rho_c = 1 - \rho_n$$

In the metallic phase $\rho_n = 1$ and $\rho_c = 0$; with the development of the CDW gap, $\rho_c \sim A^2$ grows reaching the equilibrium $\rho_c = 1$ at $T = 0$ when $\rho_n = 0$.

To proceed with the numerical solutions to the above equations, we must add, as before, the boundary and initial conditions. The initial conditions are commonly chosen as

$$n(0,x) = 0,\ \Phi(0,x) = 0\,,\ \phi(0,x) = 0\,,\ A(0,x) = 1 \tag{15}$$

Boundary conditions for A can be chosen as either $A = 1$ or $\vec{v}_\perp \nabla A = 0$ with $\vec{v}_\perp$ being a normal to the boundary. For the function $F(n, A)$ we use the simplest expression:

$$F(n,A) = \frac{n^2}{2N_F} + \frac{1}{2}A^2(-\tau + (a\xi_0 n)^2)\Delta_0^2 N_F + A^4\frac{b}{4}\Delta_0^2 N_F \tag{16}$$

where the coefficients $a, b \sim 1$. Here, the first term comes from the normal metal, and the two other terms present the Landau-type expansion in A at the proximity $\tau = 1 - T/T_c \ll 1$ of the nominal transition temperature T_c. The actual transition is shifted locally by the presence of normal carriers with a critical concentration of $n_{cr} = 1/(a\xi_0)$ where $\xi = \xi_0\tau^{1/2}$ is the correlation length. Thus, we obtain

$$\frac{\partial F}{\partial A} = A\Delta_0^2 N_F(-\tau + (a\xi_0 n)^2 + bA^2) \tag{17}$$

$$\zeta = \left(\frac{n}{N_F}\right)^2 (1 + (aA)^2\,,\ \rho_n = \frac{1}{1 + (aA)^2}\,,\ \rho_c = \frac{(aA)^2}{1 + (aA)^2}.$$

Our numerical results presented below demonstrate the possibility of vortex creation when the applied electric field or monitored current are above their thresholds. Figure 2 illustrates a one-dimensional regime corresponding to experimentally typical whisker crystals. We observe the nodes of the DW amplitude $A(x,t)$ located at both sides of the sample; they correspond periodical in time in the phase slip processes. Figure 3 illustrates a two-dimensional regime in a film with an electric field applied transversely to the CDW displacements. The achieved appearance of the static array of dislocations might be relevant to the experimentally observed step-wise reconstruction of mesa-junctions [54].

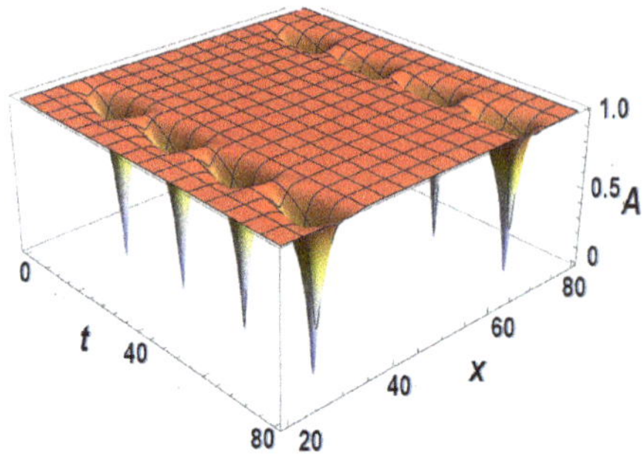

Figure 2. A sequences of phase slip events near the two sample boundaries. They are seen as sets of nodes of the DW amplitude $A(x,t)$

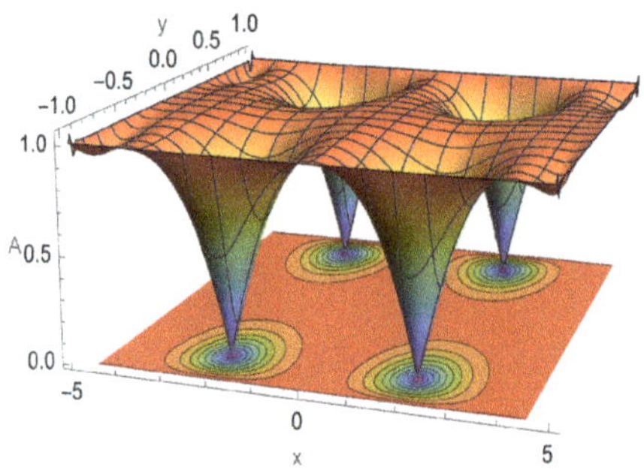

Figure 3. A fragment of a static array of dislocations appearing near sample boundaries under an electric field applied transversely to direction x of DW displacements.

4. CDW Sliding and Vorticity Under a High Magnetic Field

Commonly, the sliding CDW current is driven by the electric field E_x (in direction x of the chains), which penetration is mediated by the normal current density j_{nx} in the same direction. The sliding j_c and normal j_{nx} currents flow in the same direction x experiencing a gradual conversion towards the equilibrium. The normal current dominates at a junction; it gradually fades giving rise to the collective current until the equilibrium is reached for the bulk stationary regime [35,36].

New astonishing circumstances have appeared with experiments on mesa-junctions under a high magnetic field (HMF) [41,55]. It was found that the CDW can be driven by the Hall voltage V_H along the chains' direction x, generated by a normal current j_{ny} running in a transverse interchain direction y. Moreover, the collective and normal longitudinal j_{nx} currents now move in opposite directions; furthermore, they are completely compensated so that the total current is zero. The zero-sum loop of the collective current, exactly compensated for by the counter current of normal electrons, promotes their annihilation at the boundaries of the open circuit design, enacted by periodic phase slip processes seen as spontaneous coherent oscillations.

Following the experimental geometry, the Hall bar was formed across the CDW chains; the magnetic field and bias normal current were applied orthogonally in the interchain directions z and y correspondingly, producing the Hall voltage V_x in the direction $\hat{x}$ of the chains with a short length L_x. According to experiments, the magnetic field must be sufficiently high such that the carriers are quantized and only occupy the lowest Landau levels (LL). This level can acquire as much as $N_{LL} = 1/(2\pi\lambda^2)$ electrons per unit area where $\lambda = \sqrt{(\hbar c/eB)}$ is the magnetic length, so the LL filling per chain is $\nu = 2\pi\lambda^2\bar{n}_1/d_y, \nu \leq 1$ where $\bar{n}_1$ is the mean concentration of normal carriers with a local value of $n_1(x,t)$. In this regime of the extreme Landau quantization, the density is determined locally by the electric potential $\Phi(\vec{r})$. The profile for the density of carriers per chain is $n_1 = f_F(\Phi/T)\bar{n}_1/v$ where $f_F(z)$ is the Fermi distribution function.

There are basic local relations

$$n_c(x,t) = A^2\partial_x\phi(x,t)/\pi \ , \ j_c = -A^2\partial_t\phi(x,t)/\pi \tag{18}$$

$$n_1(x,t) + n_c(x,t) \equiv \bar{n}_1 f_F(\Phi/T)/v + \partial_x\phi/\pi = \bar{n}_1 \tag{19}$$

Here, the relations (18) are common to the conventional GL picture in Section 2.1; they provide the commonly used definitions of the collective density and current; n_c and n_1 are coupled by Equation (19) in the local electroneutrality approximation which is the limit $r_0 \to 0$ of the Poisson Equation (7).

The entangled evolution of $\phi(x,t)$ and $A(x,t)$ are described by the basic time-dependent GL equations:

$$\gamma j_c(x,t) + \partial_x n_c(x,t) - N_F\partial_x\Phi(x,t) + F_p(\phi) = 0 \tag{20}$$

$$\gamma_A\partial_t A(x,t) - \xi^2\partial_x^2 A(x,t) + \xi^2(\partial_x\phi(x,t))^2 A(x,t) - A(x,t) + A(x,t)^3 = 0$$

$$\Phi(x,t) = T\ln\left(\frac{n_1/v - n_1 + n_c(x,t)}{n_1 - n_c(x,t)}\right) \tag{21}$$

where Equation (21) is the resolution for Φ as a function of n_c of Equation (19), and $F_p(\phi)$ is the pinning force which disappears after the onset of sliding. These equations were solved for a sample with a width of l with boundary conditions $A(0,t) = A(l,t) = 1$; boundary values of the potential Φ were turned on, in view of the unique relation (21), via $n_c(0,t)$ and $n_c(l,t)$, following smooth switching $\propto \tanh(t/\tau)$.

The regime of depinned sliding with $F_p(\phi) = 0$ is illustrated in Figure 4. This shows a generated sequence of periodic phase slips where the CDW phase acquires 2π steps Figure 4, Left panel). With phase slips as the space-time vortices (Figure 4, Right panel), the amplitude $A(x,t)$ passes through zero at their cores. With our parameters, it is not completely restored between the nodes, so a persistent dip is formed from which $A(x,t)$ periodically shoots to zero (Figure 4, Center panel). This figure explains the generation of the experimentally observed coherent oscillations [41] which frequency is proportional to the bias voltage along the Hall bar.

Consider the depinning regime of CDW sliding when the driving force (the Hall voltage) just exceeds the threshold [41]. We use the same set of equations with a necessary simplification to fix the CDW amplitude $A(x,t) \equiv 1$ in the bulk. We model the pinning force $F_p(\phi)$ in such a way that it starts with the linear law $F_p(\phi) \propto \phi$ and then sharply drops towards zero at $\phi > \phi_p$, where $\phi_p \lesssim \pi$ is the threshold phase displacement of depinning:

$$F_p(\phi) = \frac{\phi/(\pi N_F L_p^2)}{1 + \exp(k(\phi - \phi_p))} \tag{22}$$

Here L_p is the pinning length which value can be taken from the experiments, and $k \gg 1$ is a modelling parameter. The space-temporal evolution of the amplitude $A(x,t)$ and the phase $\phi(x,t)$ are presented in Figure 5. The modelling reveals the formation of depinning propagating fronts created at the sample boundaries. The fronts collide, first forming a short peak of the phase velocity $\partial_t\phi/\pi$; subsequently, the fronts annihilate giving rise to the sliding regime at still inhomogeneous distributions of the electric field and particle density.

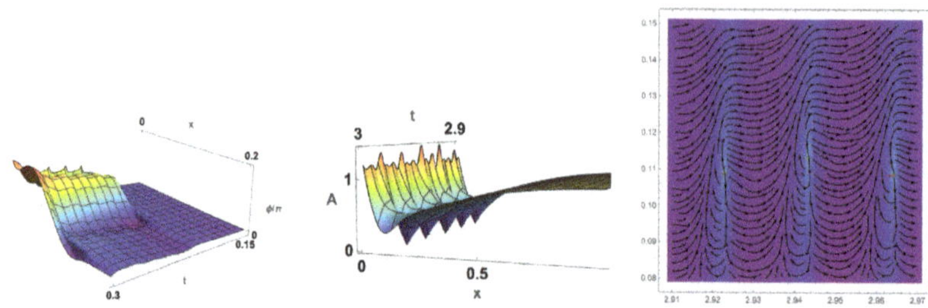

Figure 4. Plots of the spatio-temporal evolution from modelling the periodic phase slips: Left panel: the 3D plot of the phase $\phi(t,x)/\pi$); Center panel: the CDW amplitude $A(t,x)$; Right panel: streamlines on the amplitude density plot for the vector $\{u(x,t), v(x,t)\}$. A sequence of three phase slip events is observed with the amplitude nodes located near $x \approx 0.11$.

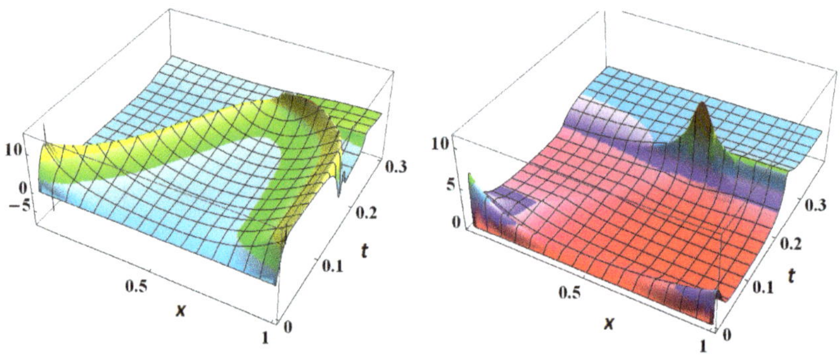

Figure 5. Evolution of the depinning process: Left panel: the self-consistent driving electric field $F(x,t) = -\partial_x\Phi(x,t)$ with $\Phi(x,t)$ given by Equation (21); Right panel: the phase velocity $\partial_t\phi$.

We summarize the content of this Section as follows. Application of a high magnetic field, normal carriers in residual pockets left after CDW formation can be brought to a quantum state at low Landau levels. When forced to carry a transverse current, the carriers generate a Hall electric field in the chains' direction which brings the CDW background to the collective sliding. Ensuring the Hall geometry has no source/drain junctions in the sliding direction, the collective current enforces the on-chain component of the normal current, exactly compensating for the two counter currents. The annihilation of the counter currents takes place near the boundaries via the space-time vortices of the CDW phase around the enforced amplitude nodes resulting in a regular sequence of the phase slips. For the first time, to our knowledge, the simplicity of Equations (due to the local relation (21) of the electric potential and the carriers' concentration) allows to model the depinning process when the depinning fronts are nucleated at the edges and develop into the bulk with their subsequent annihilation.

5. Combined Topological Defects in Spin-Density Waves

Systems with spontaneous symmetry breaking with respect to a complex multi-component order parameter may offer an intriguing possibility for non-integer vorticities in subspaces of their degeneracy (see [56]). Fractional vortices have been investigated in the superfluid helium A ([57,58], triplet superconductors [59,60], FFLO phase [61,62] and Bose-condensate of polaritons [63]. Following the suggestion of a fractionalization of the spin vortex in an anti-ferromagnetic crystal in presence of a frozen-in dislocation [64], the regular presence of combined topological defects have been suggested [65] for a sliding SDW state which is explored below.

5.1. Half-Integer Complexes of Vortices and Dislocations in SDWs.

Contrary to conventional 2π vortices created in CDWs [65], in SDWs the complex topological objects can appear as half-integer π-phase vortices combined with half-integer spin rotation. The SDW order parameter $\vec{\eta}_{SDW}$ (1) allows for three options of its self-mapping $\vec{\eta}_{SDW} \to \vec{\eta}_{SDW}$ which are determined by partial vorticities ν_ϕ and ν_m in the SDW phase and its magnetization direction.

(i) The phase dislocation $D_{\phi 2\pi}$, i.e., the pure phase ϕ vortex $V_{\phi 2\pi}$, while $\vec{m}$ stays intact: $\phi \to \phi + 2\pi \equiv \phi$ and $\vec{m} \to \vec{m}$, where $\nu_\phi = 1$ and $\nu_m = 0$;

(ii) The $\vec{m}$ - vortex $V_{\vec{m}2\pi}$, where $\vec{m}$ performs the full rotation while the phase ϕ stays intact: $\vec{m} \to \mathcal{O}_{2\pi}\vec{m} \equiv \vec{m}$ and $\phi \to \phi$, where $\nu_\phi = 0$ and $\nu_m = 1$;

(iii) The combined object $V_{\phi\pi} \otimes V_{\vec{m}\pi}$: $\phi \to \phi \pm \pi$ and $\vec{m} \to \mathcal{O}_\pi \vec{m} = -\vec{m}$, where $\nu_\phi = \pm 1/2$ and $\nu_m = \pm 1/2$. In this case, the two factors of the SDW order parameter $\vec{\eta}_{SDW} = \vec{m}\exp(i\phi)$ change the sign together, but their product leaves η_{SDW} constant.

The distribution of $\vec{m}$ near the combined half-integer vortex is illustrated schematically in Figure 6.

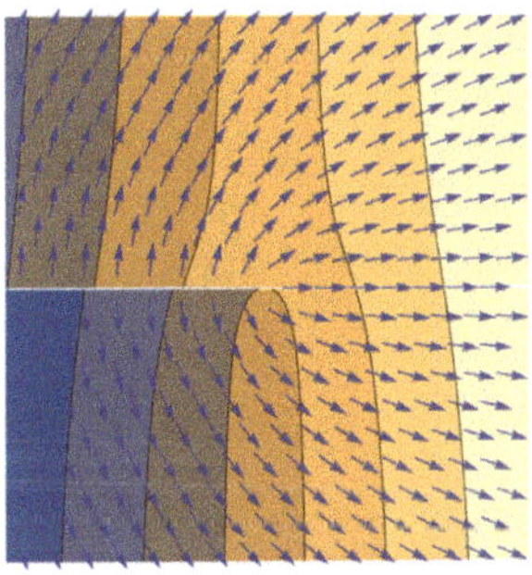

Figure 6. Configurations of the half-integer combined vortex in the SDW: vector plot of the magnetization $\vec{m}$ on top of the contours of the positional phase $Qx + \phi(x)$. The white line drawn to the left from the defect centre carries no physical singularity: the visible discontinuity in the direction of $\vec{m}$ is compensated by the π jump in ϕ.

The energy density in the SDW encompasses two parts: $w_{SDW} = w_{chrg}\{\phi\} + w_{spin}\{\vec{m}\}$. The charged part $w_{chrg}\{\phi\}$ originating from the phase displacements is common to both CDW and SDW; it is given by Equation (10). The magnetic part $w_{spin}\{\vec{m}\}$ has the form

$$w_{spin}\{\vec{m}\} = \frac{\hbar v_F}{4\pi}\left[\alpha_\parallel(\partial_x\vec{m})^2 + \alpha_\perp\left((\partial_y\vec{m})^2 + (\partial_z\vec{m})^2\right) + w_{ani}\right] \tag{23}$$

where the moduli $\alpha_\parallel, \alpha_\perp$ characterizes the rotation of the staggered magnetization with the unit vector $\vec{m}$; their values are similar to the bare moduli $C_\parallel^0, C_\perp$ of the phase displacements taken without Coulomb interactions in (10). w_{ani} is the spin anisotropy energy; its effects on the structure of combined topological defects are discussed in the next section.

Since the energies of the vortices depend on their winding numbers as $\propto \nu_{\phi,m}^2$, the common tendency is to reduce $\nu_{\phi,m}$ to the lowest permissible values. A comparison of energies of the pure phase vortex $V_{\phi 2\pi}$ (phase dislocations $D_{\phi 2\pi}$) and combined objects $V_{\phi\pi} \otimes V_{\vec{m}\pi}$ must account for the charge conservation, thus protecting the total phase vorticity $\sum \nu_\phi$. The total magnetic vorticity $\sum \nu_m$ does not need to be conserved, but should be kept at zero to avoid logarithmic energy divergence at large distances from a non-compensated pair of vortices. The energy divergence of phase dislocations is not a limitation since it is compensated for by the driving electric field. Therefore, it is tempting to consider the spitting of one $V_{\phi 2\pi}$ into two combined objects $V_{\phi\pi} \otimes V_{\pm\vec{m}\pi}$ which follows a decomposition path of a phase vortex into two pairs of half-vortices:

$$V_{\phi 2\pi} \Rightarrow \{V_{\phi\pi} \otimes V_{\vec{m}\pi}\} + \{V_{\phi\pi} \otimes V_{-\vec{m}\pi}\}$$

If the energy parameters for phase and magnetic perturbations, $\alpha_{\|}, \alpha_{\perp}$ and $C_{\|}^{0}$, $C_{\perp}$, are comparable as it takes place near to T_c when free carriers screen Coulomb interactions at shortest distances, the consequent dissociation cost is zero: $\nu_{2\pi}^2 = 1 \Rightarrow 4\nu_{\pi}^2 = 1$ and the result depends on a tiny balance of coefficients. Contrarily, at low T the Coulomb energy of the phase counterpart dominates; therefore, the magnetic vortex energy can be neglected. In this case about a half of the energy is gained by the decomposition of $(\nu_{2\pi} = 1)^2 \Rightarrow 2(\nu_{\pi} = 1/2)^2 = 1/2$ which cannot be rejected.

The energy of the pure spin vortex $V_{\vec{m}2\pi}$ is estimated as $W_{\vec{m}2\pi} \sim T_c \ln(Y/a_{\perp})$ at all transverse distances Y among the vortex lines. As demonstrated in Section 2.2, for a pure phase vortex such a form of energy persists for the screened Coulomb interaction at $Y \gg r_{scr}$ where $W_{\phi 2\pi} \sim T_c(r_{scr}/r_0)\ln(Y/r_{scr})$; at low T it is greatly enhanced as r_{scr}/r_0 with respect to $W_{\vec{m}2\pi}$. However, except when in close proximity to T_c, the energies of he initial connfiguration $V_{\phi 2\pi}$ and the $V_{\phi\pi}$ counterpart of the split configuration are highly enhanced. The energy gain is more drastic at low T at intermediate $Y < r_0$ which is discuss below in Section 5.2 in conjunction with effects of magnetic anisotropy.

Our numerical calculations confirm the spontaneous splitting of the integer 2π phase vortex $V_{\phi 2\pi}$ (Figure 7a) into two half-integer combined vortices $\{V_{\phi\pi} \otimes V_{\vec{m}\pi}, V_{\phi\pi} \otimes V_{-\vec{m}\pi}\}$ (Figure 7a,b).

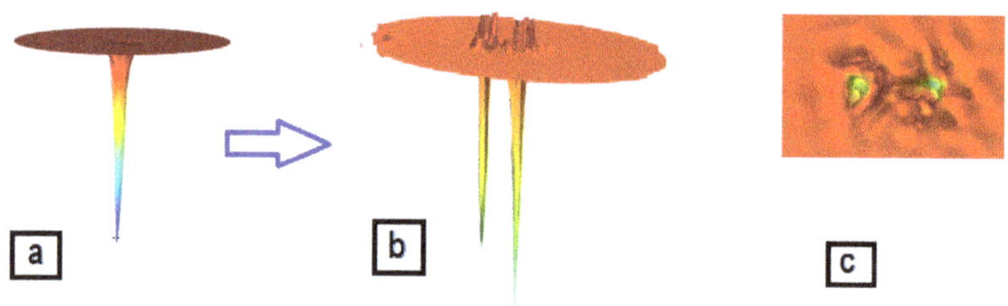

Figure 7. Calculated profiles of the amplitude $A(x,y)$ for the spontaneous splitting of the bare phase vortex (**a**) into a pair of combined, phase plus magnetization, half-integer vortices (**b** and **c** for the 3D and density plots, respectively).

5.2. *Effects of the Spin Anisotropy*

There are two cases of the spin anisotropy which still allow for a free rotation of spins, thus validating the above conclusions: (i) a full isotropy or an "easy plane" anisotropy; (ii) an "easy axis" anisotropy under a magnetic field exceeding the spin-flop threshold $H > H_{sf} \sim 1T$ (the momentum $\vec{m}$ is tilted and a free rotation of the hard plane projection is permitted). However, the known SDW crystals are of low-symmetry, possessing the spin anisotropy in all three directions. Its presence prevents the long-distance rotation of $\vec{m}$ which results in a confinement of the half-integer vortices in pairs connected by a magnetic domain wall. If the anisotropy is weak, it will not affect the arrangement in the vicinity of a vortex, but at large distances the rotation of spins is concentrated within a Neel domain wall. This forms a string (a plane in 3D) which attempts to confine the two combined objects as it is illustrated in Figure 8.

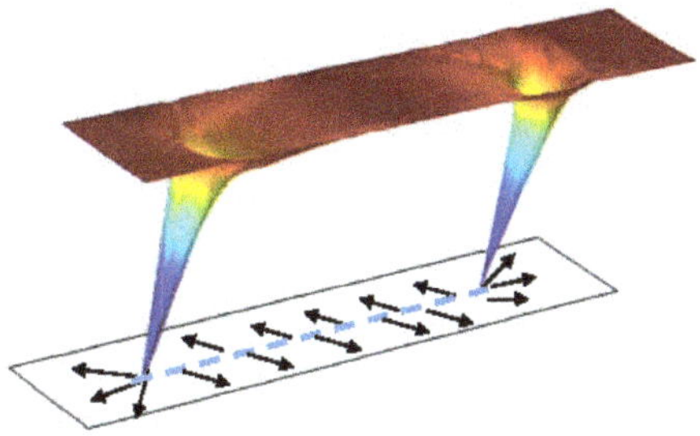

Figure 8. Two half-integer $\vec{m}$ vortices connected by a Neel domain wall.

The combined objects are composed of the spin and charge semi-vortices. The confinement, due the spin anisotropy, competes with the repulsion, due to Coulomb forces. We now recall (Section 2.2) that the unscreened Coulomb contribution to the vortex energy brings about a linear law instead of the standard logarithmic law. The total energy gain for the pair of combined vortices with respect to the normal phase vortex will be $E_C N/2$, where $E_C \sim T_c a_\perp / r_0$ and $N = Y/a_\perp$ is the number of chains between the combined vortices. The repulsive anti-confinement energy overcomes the energy lost due to domain wall formation $W_{wall} = w^A N$, as both possess a similar N dependence. Commonly, $w^A \sim 1K < E_C \sim 10^1 K$ is much smaller than E_C, hence the resulting interaction between the two semi-vortices is repulsive. We see that within the screening length $r < r_{scr}$ both the anisotropy and Coulomb energies grow linearly with distance, with the latter dominating quantitatively. Hence the anisotropy binding energy is overcome by Coulomb repulsion and thus no confinement.

At distances exceeding the screening length $r > r_{scr}$, the growth of the Coulomb energy slows down to the conventional, while quantitatively enhanced, logarithmic law (Section 2.2). However, the spin anisotropy energy W^A_{spin} keeps growing linearly. The total energy gain of the two combined vortices with respect to the dislocation becomes

$$W = -E_0 \ln N + w^A N \,,\; E_0 \sim T_c r_{scr}/r_0 = T_c/\sqrt{\rho_n}$$

As this dependence on N is non-monotonous, there exists an equilibrium distance between the two semi-vortices $N_{eq} \sim E_0/w^A \propto 10^1/\sqrt{\rho_n}$. With the screening being frozen out at low temperatures when $\rho_n \to 0$, the string length may pass through the whole sample width, typically ~ 1 µm, already at experimentally accessibly low temperatures

5.3. Combined Topological Defects and Narrow Band Noise Generation

A universal signature of DW sliding is the generation of a coherent periodic unharmonic voltage signal, the so called "narrow band noise" (NBN) (see [23,66]). In CDWs, its frequency f_{CDW} is usually proportional to the collective j_{CDW} and, remarkably, their ratio universally corresponds to the intuitively clear definitions

$$j_{CDW} = \frac{e}{\pi}\frac{\partial \phi}{\partial t} \,,\; f_{CDW} = \frac{1}{2\pi}\frac{\partial \phi}{\partial t} \qquad (24)$$

The resulting ratio $R_{CDW} = f_{CDW}/j_{CDW} = 1/2$ corresponds to the proliferation of two electrons when the CDW is displaced by its wave length λ. In CDWs the NBN generation was measured directly and the ratio of $1/2$ was definitively confirmed. In SDWs the NBN was observed less precision with a variable ratio $R_{SDW} = ef_{SDW}/j_{SDW}$ (usually small because of the inhomogeneous distribution of currents). Nevertheless, the linear law $R_{SDW} = cnst$ was confirmed.

Up till now there is no consensus on the origin of the NBN, with two competing views.

(i) The first and largely accepted was the wash-board frequency (WBF) model which suggests extrinsic NBN generation by DW sliding through the periodic host lattice or its defects [23]. Within the WBF model, the NBN frequency in SDWs is doubled with respect to CDWs. This comes from differences in the oscillating densities of the charge or lattice displacements:

$$\rho_{CDW} \sim \eta_{CDW} = A\cos[Qx + \phi]\ ,\ \phi = 2\pi f t$$

$$\rho_{SDW} \sim \delta(\vec{\eta}_{SDW})^2 \Rightarrow A^2\cos[2Qx + 2\phi]$$

because of the disparity of $\vec{\eta}_{SDW}$ with respect to the time reversal. Within the WBF model, the NBN frequency should be doubled for SDWs with respect to CDWs.

(ii) The phase slip generation (PSG) model assumes that the NBN is generated by phase slip processes created periodically near injection contacts [33,34,48–50]. This mechanism looks plausible since the phase slips are already necessary for the current conversion at the contacts, with some concerns on their regularity. Nevertheless, this regularity is clearly reproduced in our modelling, as demonstrated in Sections 2.1 and 4, in accordance with the experimentally confirmed high coherence of the NBN signal.

(iii) An intermediate scenario can be envisaged for the DW which slides in the bulk, while being pinned at the sample surface. With the coupling energy $W_{b-s} = W_0\cos(\phi_{bulk} - \phi_{surface})$, the proliferation of ϕ_{bulk} at constant $\phi_{surface}$ would provide the necessary WBF. This separation of moving and pinned layers calls for the successive creation of dislocations, thus constructing a bridge to the PSG scenario. This scenario is also successfully covered within our modelling.

While for CDWs all three scenario give the same ratio $R_{CDW} = 1/2$, for SDWs an intriguing difference was discussed (see [67,68]): $R_{SDW} = 1/2$ for the PSG mechanisms in (ii) and (iii) and $R_{SDW} = 1$ for (i) of the WBF one. Intriguingly, our prediction for chimera defects with combined half-integer vorticities indicates the possibility of $R_{SDW} = 1/2$, even for the PSG mechanisms (ii) and (iii). Moreover, the ratio can switch between $1/2$ and 1 depending on the temperature. Indeed, the chimeras are energetically favourable at low temperatures in the absence of screening carriers, when the Coulomb energy dominates the phase vorticity. The π vorticity of the phase requires a double slip rate to promote the same collective current. However, near T_c the energies of the magnetic and phase vorticities are comparable, so the splitting into half-integer objects is not strongly profitable bringing back R_{SDW} from 1 to $1/2$. As we have explained in the previous sections, the magnetic anisotropy can confine the half-integer object into loose integer pairs, thus re-tuning R_{SDW} to $1/2$ for sufficiently narrow samples.

To summarise this section, the sliding SDW generates combined topological objects, where the DW displacement and spin rotations are topologically coupled. These defects lower the Coulomb energy of the conventional phase vortex and are energetically stable. Their formation also leads to a twice reduction in the NBN frequency, which remains an important debatable topic.

6. Conclusions

In the science of electronic crystals, ample research has been devoted to the study of stationary or transient states with static and dynamic topologically non-trivial configurations, such as electronic vortices (dislocations), instantons (phase slip centres), and microscopic soliton complexes. Having been inspired by extensive experiments, we have studied equations for dynamics involving topological objects in density waves under various circumstances. Our models include multiple fields in their mutual non-linear interactions: the phase and amplitude of CDWs, the spin polarization for SDWs, electric field, and distributions of current and density of normal carriers. The numerical solutions to the resulting equations illustrate the emergence of vortices, their turbulent evolution, and the final formation of steady arrays or stationary oscillations.

We have studied several experimentally significant situations, such as proliferations of space vortices under the field effect of a transverse voltage and a coherent sequence of space-time vortices, or phase slips, under longitudinal driving. Furthermore, the sliding driven by the Hall field generated by quantized electrons under a high magnetic field was modelled, revealing an interesting regime of exactly compensated collective and normal countercurrents; their annihilation proceeds via a regular sequence of phase slip processes.

SDWs add an additional richness by combining the common vorticity of the phase with the particular vorticity of modulated magnetization. This gives rise to interesting topological objects with normally prohibited half-integer winding numbers. The standard dislocation splits into two combined topological objects which, in the presence of a weak spin anisotropy, are connected by a string whichconfining tension can be overrun by the Coulomb energy of the phase component. The formation of such combined topological objects results in a two-fold reduction in the NBN frequency, which still remains an essential disputed question.

The combined topological defects link our specific problems with the general issues reported for quantum solids and liquids, particle physics, and cosmology.

Author Contributions: Conceptualization, N.K. and S.B.; methodology, N.K. and S.B.; formal analysis, N.K. and S.B.; writing—original draft preparation, N.K. and S.B.; writing—review and editing, N.K. and S.B. All authors have read and agreed to the published version of the manuscript.

Funding: This research received no external funding

Data Availability Statement: Not applicable.

Conflicts of Interest: The authors declare no conflict of interest.

References

1. Mermin, N.D. The topological theory of defects in ordered media. *Rev. Mod. Phys.* **1979**, *51*, 591–648. [CrossRef]
2. Mineev, V.P. *Topologically Stable Defects and Solitons in Ordered Media*; Harwood Academic Publisher: New York, NY, USA, 1998
3. Bunkov, Y.M.; Godfrin, H. (Eds.) *Topological Defects and the Non-Equilibrium Dynamics of Symmetry-Breaking Phase Transitions*; NATO ASI Series; Kluwer Academic Publishers: Amsterdam, The Netherlands, 2000.
4. Skyrme, T.H.R. A non-linear field theory. *Proc. R. Soc. Lond. Ser. A* **1961**, *260*, 127–138.
5. Hindmarsh, M.W.; Kibble, T.B.W. Cosmic strings. *Rep. Prog. Phys.* **1995**, *58*, 477–561. [CrossRef]
6. Vilenkin, A.; Shellard, E.P.S. *Cosmic Strings and Other Topological Defects*; Cambridge University Press: Cambridge, UK, 2000.
7. Schiffer, P.; Osheroff, D.D.; Leggett, A.J. Nucleation of the ab transition in superfluid ^{3}He: Experimental and theoretical considerations *Progr. Low Temp. Phys.* **1995**, *14*, 159–211.
8. Volovik, G.E. *The Universe in a Helium Droplet*; Clarendon Press: Oxford, UK, 2003
9. Eltsov, V.B.; Krusius, M.; Volovik, G.E. Vortex formation and dynamics in superfluid ^{3}He and analogies in quantum field theory. *Prog. Low Temp. Phys.* **2005**, *15*, 1–137.
10. Berloff, N.G.; Svistunov, B.V. Scenario of strongly nonequilibrated Bose-Einstein condensation. *Phys. Rev. A* **2002**, *66*, 013603–013610. [CrossRef]
11. Berloff, N.G.; Keeling, J. Universality in Modelling Non-equilibrium Pattern Formation in Polariton Condensates. In *Physics of Quantum Fluids: New Trends and Hot Topics in Atomic and Polariton Condensates*; Springer Series 177; Bramati, A., Modugno, M., Eds.; Springer: Berlin/Heidelberg, Germany, 2013; pp. 19–38.
12. Neely, T.W.; Scherer, D.R.; Bradley, A.S.; Davis, M.J.; Anderson, B.P. Spontaneous vortices in the formation of Bose-Einstein condensates. *Nature* **2008**, *455*, 948–951.
13. Chuang, I.; Durrer, R.; Turok, N.; Yurke, B. Cosmology in the laboratory: Defect dynamics in liquid crystals. *Science* **1991**, *251*, 1336–1342. [CrossRef]
14. Bowick, M.J.; Chander, L.; Sciff, E.A.; Srivastava, A.M. The Cosmological Kibble Mechanism in the Laboratory: String Formation in Liquid Crystals. *Science* **1994**, *263*, 943–945. [CrossRef]
15. Kranjec, A.; Karpov, P.; Vaskivskyi, Y.; Vodeb, J.; Gerasimenko, Y.; Mihailovic, Y. Electronic Dislocation Dynamics in Metastable Wigner Crystal States. *Symmetry* **2022**, *14*, 926. [CrossRef]
16. Kopnin, N. *Theory of Nonequilibrium Superconductivity*; Oxford University Press: Oxford, UK, 2001.
17. Landau, L.D.; Pitaevskii, L.P.; Kosevich, A.M.; Lifshitz, E.M. *Theory of Elasticity*; Pergamon: Oxford, UK, 1984.
18. Kosevich, A.M.; Ivanov, B.A.; Kovalev, A.S. Magnetic Solitons. *Phys. Rep.* **1990**, *194*, 117–238. [CrossRef]
19. Seidel, J. (Ed.) *Topological Structures in Ferroic Materials: Domain Walls, Vortices and Skyrmions*; Springer Series in Materials Science; Springer: Berlin/Heidelberg, Germany, 2016.
20. Seidel, J. Scanning Probe Microscopy Investigation of Topological Defects. *Symmetry* **2022**, *14*, 1098 [CrossRef]
21. Monceau, P. Electronic crystals: An experimental overview. *Adv. Phys.* **2012**, *61*, 325–581. [CrossRef]
22. Available online: https://www.sciencedirect.com/journal/physica-b-condensed-matter/vol/460/suppl/C (accessed on 4 February 2023).

23. Grüner, G. *Density Waves in Solids*; CRC Press: Boca Raton, FL, USA, 2018.
24. Gor'kov, L.; Grüner, G. (Eds.) *Charge Density Waves in Solids*; Elsevier Sci. Publ.: Amsterdam, The Netherlands, 1990.
25. Grüner, G. The dynamics of spin-density waves. *Rev. Mod. Phys.* **1994**, *66*, 1–24. [CrossRef]
26. Bardeen, J. Superconductivity and Other Macroscopic Quantum Phenomena. *Phys. Today* **1990**, *43*, 25–33. [CrossRef]
27. Ivlev, B.I.; Kopnin, N.B. Electric currents and resistive states in thin superconductors. *Adv. Phys.* **1984**, *33*, 47–114. [CrossRef]
28. Gor'kov, L.P. Phase slipping processes and the generation of narrow bad oscillationsby charge desity waves. *Charge Density Waves in Solids*; Gor'kov, L., Grüner, G., Eds.; Elsevier Sci. Publ.: Amsterdam, The Netherlands, 1990; pp. 403–423.
29. Feinberg, D.; Friedel, J. Elastic and plastic deformations of charge density waves. *J. Phys.* **1988**, *49*, 485–496. [CrossRef]
30. Feinberg, D.; Friedel, J. Imperfections of Charge Density Waves in Blue Bronzes *Low-Dimensional Electronic Properties of Molybdenum Bronzes and Oxides*; Schlenker, C., Ed.; Kluwer Ac.: Dordrecht, The Netherlands, 1989; pp. 407–448.
31. Brazovskii, S.I. Matveenko, Charged density wave structure near contacts. *Sov. Phys. JETP* **1992**, *74*, 864–870.
32. Hayashi, M.; Yoshioka, H. Topological Dislocations and Mixed State of Charge Density Waves. *Phys. Rev. Lett.* **1996**, *77*, 3403–3407. [CrossRef]
33. Ong, N.P.; Varma, G.; Maki, K. Vortex-Array Model for Charge-Density-Wave Conduction Noise. *Phys. Rev. Lett.* **1984**, *52*, 663–667. [CrossRef]
34. Ong, N.P.; Maki, K. Generation of charge-density-wave conduction noise by moving phase vortices. *Phys. Rev. B* **1985**, *32*, 6582–6590. [CrossRef] [PubMed]
35. Lemay, S.G.; de Lind van Wijngaarden, M.C.; Adelman, T.L.; Thorne, R.E. Spatial distribution of charge-density-wave phase slip in NbSe3. *Phys. Rev. B* **1998**, *57*, 12781–12791. [CrossRef]
36. Rideau, D.; Monceau, P.; Currat, R.; Requardt, H.; Nad, F.; Lorenzo, J.E.; Brazovskii, S.; Detlefs, C.; Gruebel, G. X-ray scattering evidence for macroscopic strong pinning centers in the sliding CDW state of NbSe3. *Eur. Lett.* **2001**, *56*, 289–295. [CrossRef]
37. Brazovskii, S.; Matveenko, S. Dislocations in charge density wave crystals. *Sov. Phys. JETP* **1991**, *74*, 860–865.
38. Kirova, N.; Brazovskii, S. Scattering profile from a random distribution of dislocations in a charge density wave. *Ann. Phys.* **2022**, *447*, 169130. [CrossRef]
39. Nad', F.Y. Deformation and phase-slip in charge density waves in quasi-one-dimensional conductors. In *Charge Density Waves in Solids*; Gor'kov, L., Grüner, G., Eds.; Elsevier Sci. Publ.: Amsterdam, The Netherlands, 1990; pp. 191–221.
40. Brazovskii, S.; Kirova, N. Multi-Fluid Hydrodynamics in Charge Density Waves with Collective, Electronic, and Solitonic Densities and Currents. *JETP* **2019**, *129*, 659–668. [CrossRef]
41. Orlov, A.P.; Sinchenko, A.A.; Monceau, P.; Brazovskii, S.; Latyshev, Y.I. Hall voltage drives pulsing counter-currents of the sliding charge density wave and of quantized normal carriers at self-filled Landau levels. *NPJ Quantum Mater.* **2017**, *2*, 61. [CrossRef]
42. Brazovskii, S.; Kirova, N. Phase slips, dislocations, half-integer vortices, two-fluid hydrodynamics and the chiral anomaly in charge and spin density waves. *JETP* **2021**, *159*, 806–814.
43. Yi, T.; Kirova, N.; Brazovskii, S. Modeling of dynamics of field-induced transformations in charge density waves. *Eur. Phys. J. Spec. Top.* **2013**, *222*, 1035–1046. [CrossRef]
44. Rojo-Bravo, A.; Yi, T.; Luo, Y.; Kirova, N.; Brazovskii, S.S. Modelling of dislocations in a CDW junction: Interference of the CDW and normal carriers. *Phys. B Condens. Matter* **2015**, *460*, 16–20. [CrossRef]
45. Brazovskii, S.; Kirova, N. From chiral anomaly to two-fluid hydrodynamics for electronic vortices. *Ann. Phys.* **2019**, *403*, 184–197. [CrossRef]
46. Cross, M.; Hohenberg, P. Pattern formation outside of equilibrium. *Rev. Mod. Phys.* **1993**, *65*, 851–1112. [CrossRef]
47. Aranson, I.; Kramer, L. The world of the complex Ginzburg-Landau equation. *Rev. Mod. Phys.* **2002**, *74*, 99–143. [CrossRef]
48. Gor'kov, L.P. Boundary conditions and generation of periodic noise by a space-charge wave. *JETP Lett.* **1983**, *38*, 87–90.
49. Gor'kov, L.P. Generation of oscillations by a running charge density wave. *Sov. Phys. JETP* **1984**, *59*, 1057–1064.
50. Batistic, I.; Bjelis, A.; Gor'kov, L. Generation of the coherent pulses by the CDW-motion. Solutions of the microscopic model equations. *J. Phys.* **1984**, *45*, 1049–1059. [CrossRef]
51. Jelcié, D.; Bjelis, A.; Batistic, I. Interference effects in nonlinear charge-density-wave dynamics. *Phys. Rev. B* **1988**, *38*, 4045–4055.
52. Jelcié, D.; Bjelis, A. Nonlinear charge-density-wave dynamics in inhomogeneous conditions. *Phys. Rev. B* **1991**, *43*, 1735–1742.
53. Artemenko, S.N.; Volkov, A.F.; Kruglov, A.N. Dynamics of phase slip center of charge-density wave. *Sov. Phys. JETP* **1986**, *64*, 906–913.
54. Latyshev, Y.I.; Monceau, P.; Brazovskii, S.; Orlov, A.P.; Fournier, T. Subgap collective tunneling and its staircase structure in charge density waves. *Phys. Rev. Lett.* **2006**, *96*, 116402. [CrossRef]
55. Frolov, A.V.; Orlov, A.P.; Sinchenko, A.A.; Monceau, P. Charge density wave sliding driven by an interplay of conventional and Hall voltages in NbSe3 microbridges. *Phys. Rev.* **2019**, *100*, 245126–245131. [CrossRef]
56. Mineev, V.P. Half-quantum vortices. *Low Temp. Phys.* **2013**, *39*, 1056–1061. [CrossRef]
57. Volovik, G.E.; Mineev, V.P. Line and point singularities in superfluid He3. *JETP Lett.* **1976**, *24*, 561–563.
58. Cross, M.C.; Brinkman, W.F.J. Textural singularities in the superfluid A phase of 3He. *Low Temp. Phys.* **1977**, *27*, 683–686. [CrossRef]
59. Lukyanchuk, I.A.; Zhitomirsky, M.E. Magnetic Properties of Unconventional Superconductors. *Supercond. Rev.* **1995**, *1*, 207–250.
60. Volovik, G. Monopoles and fractional vortices in chiral superconductors. *Proc. Natl. Acad. Sci. USA* **2000**, *97*, 2431–2436. [CrossRef]
61. Radzihovsky, L.L.; Vishwanath, A. Quantum Liquid Crystals in an Imbalanced Fermi Gas: Fluctuations and Fractional Vortices in Larkin-Ovchinnikov States. *Phys. Rev. Lett.* **2009**, *103*, 010404. [CrossRef]

62. Brazovskii, S. Solitons: From Charge Density Waves to FFLO in Superconductors. *Phys. B Condens. Matter* **2009**, *404*, 482–486. [CrossRef]
63. Rubo, Y.G. Half Vortices in Exciton Polariton Condensates. *Phys. Rev. Lett.* **2007**, *99*, 106401. [CrossRef]
64. Dzyaloshinskii, I.E. Domains and dislocations in antiferromagnets. *JETP Lett.* **1977**, *25*, 98–100.
65. Kirova, N.; Brazovskii, S. Topological defects in Spin Density Waves. *J. de Phys. IV* **2000**, *10*, 183–189. [CrossRef]
66. Brown, S.E.; Zettle, A. Charge density wave current oscillations and interference effects. In *Charge Density Waves in Solids*; Gor'kov, L., Grüner, G., Eds.; Elsevier Sci. Publ.: Amsterdam, The Netherlands, 1990; pp. 223–291.
67. Basletić, M.; Biškup, N.; Korin-Hamzić, B.; Hamzić, A.; Tomić, S.; Sliding spin-density waves: Studies of conduction noise, magnetic field dependence and Hall resistivity. *Fizika A* **1999**, *8*, 293–310.
68. Kriza, G.; Quirion, G.; Traetteberg, O.; Kang, W.; Jerome, D. Shapiro interference in a spin-density-wave system. *Phys. Rev. Lett.* **1991**, *66*, 1922. [CrossRef]

MDPI AG
Grosspeteranlage 5
4052 Basel
Switzerland
Tel.: +41 61 683 77 34

Symmetry Editorial Office
E-mail: symmetry@mdpi.com
www.mdpi.com/journal/symmetry

www.ingramcontent.com/pod-product-compliance
Lightning Source LLC
Chambersburg PA
CBHW062048220926
51850CB00010B/2703
9783725855216